S0-AZE-591

CCNA 200-301 Portable Command Guide

Fifth Edition

Scott Empson

Cisco Press

CCNA 200-301 Portable Command Guide, Fifth Edition

Scott Empson

Copyright© 2020 Cisco Systems, Inc.

Published by:

Cisco Press

All rights reserved. No part of this book may be reproduced or transmitted in any form or by any means, electronic or mechanical, including photocopying, recording, or by any information storage and retrieval system, without written permission from the publisher, except for the inclusion of brief quotations in a review.

260 2024

Library of Congress Control Number: 2019951511

ISBN-13: 978-0-13-593782-2

ISBN-10: 0-13-593782-5

Warning and Disclaimer

This book is designed to provide information about the Cisco Certified Network Associate (CCNA) exam (200-301). Every effort has been made to make this book as complete and as accurate as possible, but no warranty or fitness is implied.

The information is provided on an "as is" basis. The authors, Cisco Press, and Cisco Systems, Inc. shall have neither liability nor responsibility to any person or entity with respect to any loss or damages arising from the information contained in this book or from the use of the discs or programs that may accompany it.

The opinions expressed in this book belong to the author and are not necessarily those of Cisco Systems, Inc.

Trademark Acknowledgments

All terms mentioned in this book that are known to be trademarks or service marks have been appropriately capitalized. Cisco Press or Cisco Systems, Inc., cannot attest to the accuracy of this information. Use of a term in this book should not be regarded as affecting the validity of any trademark or service mark.

Microsoft and/or its respective suppliers make no representations about the suitability of the information contained in the documents and related graphics published as part of the services for any purpose. All such documents and related graphics are provided "as is" without warranty of any kind. Microsoft and/or its respective suppliers hereby disclaim all warranties and conditions with regard to this information, including all warranties and conditions of merchantability, whether express, implied or statutory, fitness for a particular purpose, title and non-infringement. In no event shall Microsoft and/or its respective suppliers be liable for any special, indirect or consequential damages or any damages whatsoever resulting from loss of use, data or profits, whether in an action of contract, negligence or other tortious action, arising out of or in connection with the use or performance of information available from the services.

Editor-in-Chief
Mark Taub

Alliances Manager, Cisco Press
Arezou Gol

Product Line Manager
Brett Bartow

Senior Editor
James Manly

Managing Editor
Sandra Schroeder

Development Editor
Ellie Bru

Senior Project Editor
Tonya Simpson

Copy Editor
Bill McManus

Technical Editor
Rick McDonald

Editorial Assistant
Cindy Teeters

Cover Designer
Chuti Prasertsith

Composition
codeMantra

Indexer
Lisa Stumpf

Proofreader
Abigail Bass

This book is part of the Cisco Networking Academy® series from Cisco Press. The products in this series support and complement the Cisco Networking Academy curriculum. If you are using this book outside the Networking Academy, then you are not preparing with a Cisco trained and authorized Networking Academy provider.

cisco.

For more information on the Cisco Networking Academy or to locate a Networking Academy, Please visit www.netacad.com

The documents and related graphics contained herein could include technical inaccuracies or typographical errors. Changes are periodically added to the information herein. Microsoft and/or its respective suppliers may make improvements and/or changes in the product(s) and/or the program(s) described herein at any time. Partial screenshots may be viewed in full within the software version specified.

Microsoft® and Windows® are registered trademarks of the Microsoft Corporation in the U.S.A. and other countries. Screenshots and icons reprinted with permission from the Microsoft Corporation. This book is not sponsored or endorsed by or affiliated with the Microsoft Corporation.

Special Sales

For information about buying this title in bulk quantities, or for special sales opportunities (which may include electronic versions; custom cover designs; and content particular to your business, training goals, marketing focus, or branding interests), please contact our corporate sales department at corpsales@pearsoned.com or (800) 382-3419.

For government sales inquiries, please contact governmentsales@pearsoned.com.

For questions about sales outside the U.S., please contact intlcs@pearson.com.

Feedback Information

At Cisco Press, our goal is to create in-depth technical books of the highest quality and value. Each book is crafted with care and precision, undergoing rigorous development that involves the unique expertise of members from the professional technical community.

Readers' feedback is a natural continuation of this process. If you have any comments regarding how we could improve the quality of this book, or otherwise alter it to better suit your needs, you can contact us through email at feedback@ciscopress.com. Please make sure to include the book title and ISBN in your message.

We greatly appreciate your assistance.

Americas Headquarters	Asia Pacific Headquarters	Europe Headquarters
Cisco Systems, Inc.	Cisco Systems (USA) Pte. Ltd.	Cisco Systems International BV Amsterdam,
San Jose, CA	Singapore	The Netherlands

Cisco has more than 200 offices worldwide. Addresses, phone numbers, and fax numbers are listed on the Cisco Website at www.cisco.com/go/offices.

Cisco and the Cisco logo are trademarks or registered trademarks of Cisco and/or its affiliates in the U.S. and other countries. To view a list of Cisco trademarks, go to this URL: www.cisco.com/go/trademarks. Third party trademarks mentioned are the property of their respective owners. The use of the word partner does not imply a partnership relationship between Cisco and any other company. (1110R)

Contents at a Glance

Part V: Security Fundamentals

Part VI: Wireless Technologies

Part VII Appendices

Contents

Part V: Security Fundamentals

About the Author

Scott Empson is an instructor in the Department of Information Systems Technology at the Northern Alberta Institute of Technology in Edmonton, Alberta, Canada, where he has taught for more than 20 years. He teaches technical courses in Cisco routing and switching, along with courses in professional development and leadership. He has a Master of Education degree along with three undergraduate degrees: a Bachelor of Arts, with a major in English; a Bachelor of Education, again with a major in English/language arts; and a Bachelor of Applied Information Systems Technology, with a major in network management. Scott lives in Edmonton, Alberta, with his wife, Trina, and two university-attending-but-still-haven't-moved-out-yet-but-hope-to-move-out-as-soon-as-possible-after-graduation-so-Dad-can-have-the-TV-room-back children, Zachariah and Shaelyn.

About the Technical Reviewer

Rick McDonald teaches computer and networking courses at the University of Alaska Southeast in Ketchikan, Alaska. He holds a B.A. degree in English and an M.A. degree in Educational Technology from Gonzaga University in Spokane, WA. After several years in the airline industry, Rick returned to full-time teaching. Rick started in the Cisco Academy in North Carolina and taught CCNA and CCNP courses to students and was a CCNA Instructor Trainer. Previous Academy projects include authoring CCNP study guides and technical editing a previous edition of the CCNA 2 and 3 textbook. His current project is developing methods for delivering hands-on training via distance in Alaska using web conferencing and NETLAB tools.

Dedications

As always, this book is dedicated to Trina, Zach, and Shae. Now that you are older and are in university, do you even know what I do when I write these books, or are you just happy that I can afford to take you to Disney again? Or pay for your tuition. Pick one... xxxooo :)

Acknowledgments

Just as it takes many villagers to raise a child, it takes many people to create a book. Without the following, I wouldn't be able to call myself an author; my title would probably be village idiot. Therefore, I must thank:

The team at Cisco Press. Once again, you amaze me with your professionalism and the ability to make me look good. James, Ellie, Bill, Tonya, and Vaishnavi: Thank you for your continued support and belief in my little engineering journal.

To my technical reviewer, Rick: We finally get to work together! Rick was one of the first people I met when getting involved with Cisco and the Cisco Academy all those years ago (2001?). I first met you in Las Vegas at a Networkers conference. You were brilliant then, and you are brilliant now. Thanks for correcting my mistakes and making me look smarter than I really am.

A special thanks to Mary Beth Ray: You were my first editor with Cisco Press and you were with me for every step over the last 15 years. Thank you for taking a risk on me and my idea. I hope that your post-publishing career is just as exciting and rewarding as your time was with us. I bow to the divine in you. Namaste.

If you like this book, it is all because of them. Any errors in this book are all on me.

Reader Services

Register your copy at www.ciscopress.com/title/9780135937822 for convenient access to downloads, updates, and corrections as they become available. To start the registration process, go to www.ciscopress.com/register and log in or create an account*. Enter the product ISBN 9780135937822 and click Submit. When the process is complete, you will find any available bonus content under Registered Products.

*Be sure to check the box that you would like to hear from us to receive exclusive discounts on future editions of this product.

Command Syntax Conventions

The conventions used to present command syntax in this book are the same conventions used in the IOS Command Reference. The Command Reference describes these conventions as follows:

- **Boldface** indicates commands and keywords that are entered literally as shown. In actual configuration examples and output (not general command syntax), boldface indicates commands that are manually input by the user (such as a **show** command).
- *Italic* indicates arguments for which you supply actual values.
- Vertical bars (|) separate alternative, mutually exclusive elements.
- Square brackets ([ ]) indicate an optional element.
- Braces ({ }) indicate a required choice.
- Braces within brackets ([{ }]) indicate a required choice within an optional element.

Introduction

Welcome to *CCNA 200-301 Portable Command Guide*! As most of you know, Cisco has announced a complete revamp and update to its certifications. What you have here is the latest Portable Command Guide as part of these new outcomes and exams. For someone who originally thought that this book would be less than 100 pages in length and limited to the Cisco Networking Academy program for its complete audience, I am continually amazed that my little engineering journal has caught on with such a wide range of people throughout the IT community.

I have long been a fan of what I call the "engineering journal," a small notebook that can be carried around and that contains little nuggets of information—commands that you forget, the IP addressing scheme of some remote part of the network, little reminders about how to do something you only have to do once or twice a year (but is vital to the integrity and maintenance of your network). This journal has been a constant companion by my side for the past 20 years; I only teach some of these concepts every second or third year, so I constantly need to refresh commands and concepts and learn new commands and ideas as Cisco releases them. My journals are the best way for me to review because they are written in my own words (words that I can understand). At least, I had better understand them because if I can't, I have only myself to blame.

My first published engineering journal was the *CCNA Quick Command Guide*; it was organized to match the (then) order of the Cisco Networking Academy program. That book then morphed into the *Portable Command Guide*, the fifth edition of which you are reading right now. This book is my "industry" edition of the engineering journal. It contains a different logical flow to the topics, one more suited to someone working in the field. Like topics are grouped together: routing protocols, switches, troubleshooting. More complex examples are given. IPv6 has now been integrated directly into the content chapters themselves. IPv6 is not something new that can be introduced in a separate chapter; it is part of network designs all around the globe, and we need to be as comfortable with it as we are with IPv4. The popular "Create Your Own Journal" appendix is still here (blank pages for you to add in your own commands that you need in your specific job). We all recognize the fact that no network administrator's job can be so easily pigeonholed as to just working with CCNA topics; you all have your own specific jobs and duties assigned to you. That is why you will find those blank pages at the end of the book. Make this book your own; personalize it with what you need to make it more effective. This way your journal will not look like mine.

Private Addressing Used in This Book

This book uses RFC 1918 addressing throughout. Because I do not have permission to use public addresses in my examples, I have done everything with private addressing. Private addressing is perfect for use in a lab environment or in a testing situation because it works exactly like public addressing, with the exception that it cannot be routed across a public network.

Who Should Read This Book

This book is for those people preparing for the CCNA certification exam, whether through self-study, on-the-job training and practice, or study within the Cisco Networking Academy program. There are also some handy hints and tips along the way to make life a bit easier for you in this endeavor. This book is small enough that you will find it easy to carry around with you. Big, heavy textbooks might look impressive on your bookshelf in your office, but can you really carry them around with you when you are working in some server room or equipment closet somewhere?

Optional Sections

A few sections in this book have been marked as optional. These sections cover topics that are not on the CCNA certification exam, but they are valuable topics that should be known by someone at a CCNA level. Some of the optional topics might also be concepts that are covered in the Cisco Networking Academy program courses.

Organization of This Book

This book follows a logical approach to configuring a small to mid-size network. It is an approach that I give to my students when they invariably ask for some sort of outline to plan and then configure a network. Specifically, this approach is as follows:

Part I: Network Fundamentals

- **Chapter 1, "IPv4 Addressing—How It Works":** An overview of the rules of IPv4 addressing—how it works, what is it used for, and how to correctly write out an IPv4 address

- **Chapter 2, "How to Subnet IPv4 Addresses":** An overview of how to subnet, examples of subnetting (both a Class B and a Class C address), and using the binary AND operation

- **Chapter 3, "Variable Length Subnet Masking (VLSM)":** An overview of VLSM, and an example of using VLSM to make your IP plan more efficient

- **Chapter 4, "Route Summarization":** Using route summarization to make your routing updates more efficient, an example of how to summarize a network, and necessary requirements for summarizing your network

- **Chapter 5, "IPv6 Addressing—How It Works":** An overview of the rules for working with IPv6 addressing, including how it works, what is it used for, how to correctly write out an IPv6 address, and the different types of IPv6 addresses

- **Chapter 6, "Cables and Connections":** An overview of how to connect to Cisco devices, which cables to use for which interfaces, and the differences between the TIA/EIA 568A and 568B wiring standards for UTP

- **Chapter 7, "The Command-Line Interface":** How to navigate through Cisco IOS Software: editing commands, using keyboard shortcuts for commands, and using help commands

Part II: LAN Switching Technologies

- **Chapter 8, "Configuring a Switch":** Commands to configure Catalyst switches: names, passwords, IP addresses, default gateways, port speed and duplex, and static MAC addresses

- **Chapter 9, "VLANs":** Configuring static VLANs, troubleshooting VLANs, saving and deleting VLAN information, and configuring voice VLANs with and without trust

- **Chapter 10, "VLAN Trunking Protocol and Inter-VLAN Communication":** Configuring a VLAN trunk link, configuring VTP, verifying VTP, and configuring inter-VLAN communication using router-on-a-stick, subinterfaces, and SVIs

- **Chapter 11, "Spanning Tree Protocol":** Verifying STP, setting switch priorities, working with optional features, and enabling Rapid Spanning Tree

- **Chapter 12, "EtherChannel":** Creating and verifying Layer 2 and Layer 3 EtherChannel groups between switches

- **Chapter 13, "Cisco Discovery Protocol (CDP) and Link Layer Discovery Protocol (LLDP)":** Customizing and verifying both CDP and LLDP

Part III: Routing Technologies

- **Chapter 14, "Configuring a Cisco Router":** Commands needed to configure a single router: names, passwords, configuring interfaces, MOTD and login banners, IP host tables, saving and erasing your configurations

- **Chapter 15, "Static Routing":** Configuring IPv4 and IPv6 static routes in your internetwork

- **Chapter 16, "Open Shortest Path First (OSPF)":** Configuring and verifying OSPFv2 in single-area designs

Part IV: IP Services

- **Chapter 17, "DHCP":** Configuring and verifying DHCP on a Cisco IOS router, and using Cisco IP Phones with a DHCP server

- **Chapter 18, "Network Address Translation (NAT)":** Configuring and verifying NAT and PAT

- **Chapter 19, "Configuring Network Time Protocol (NTP)":** Configuring and verifying NTP, setting the local clock, and using time stamps

Part V: Security Fundamentals

- **Chapter 20, "Layer Two Security Features":** Setting passwords, configuring switch port security, using static and sticky MAC addresses, configuring and verifying DHCP snooping, and configuring and verifying Dynamic ARP Inspection (DAI)

- **Chapter 21, "Managing Traffic Using Access Control Lists (ACLs)":** Configuring standard ACLs, using wildcard masks, creating extended ACLs, creating named ACLs, using sequence numbers in named ACLs, verifying and removing ACLs, and configuring and verifying IPv6 ACLs

- **Chapter 22, "Device Monitoring and Hardening":** Device monitoring, backups, logging and the use of syslog, syslog message formats, configuring and encrypting passwords, configuring and verifying SSH, restricting virtual terminal access, and disabling unused services

Part VI: Wireless Technologies

- **Chapter 23, "Configuring and Securing a WLAN AP":** The initial setup for a Wireless LAN Controller, monitoring a WLC, configuring VLANs, DHCP, WLAN, RADIUS servers, other management options, and security on a WLC

Part VII: Appendices

- **Appendix A, "How to Count in Decimal, Binary, and Hexadecimal":** A refresher on how to count in decimal, and using those rules to count in binary and hexadecimal

- **Appendix B, "How to Convert Between Number Systems":** Rules to follow when converting between the three numbering systems used most often in IT: decimal, binary, and hexadecimal

- **Appendix C, "Binary/Hex/Decimal Conversion Chart":** A chart showing numbers 0 through 255 in the three numbering systems of binary, hexadecimal, and decimal

- **Appendix D, "Create Your Own Journal Here":** Some blank pages for you to add in your own specific commands that might not be in this book

Did I Miss Anything?

I am always interested to hear how my students, and now readers of my books, do on both certification exams and future studies. If you would like to contact me and let me know how this book helped you in your certification goals, please do so. Did I miss anything? Let me know. Contact me at PCG@empson.ca or through the Cisco Press website, http://www.ciscopress.com.

Figure Credits

Figure 6-3, screenshot of PC Settings © Microsoft, 2019.

Figure 23-7, 23 Logging into the WLC Screenshot of Logging into © Microsoft, 2019.

Figure 23-15, screenshot of Interface Address © Microsoft, 2019.

Figure 23-16, screenshot of Interface Address © Microsoft, 2019.

Figure 23-17, screenshot of Success ping message © Microsoft, 2019.

Figure 23-24, screenshot of Saving configuration © Microsoft, 2019.

CHAPTER 1

IPv4 Addressing—How It Works

This chapter provides information concerning the following topics:

- What are IPv4 addresses used for?
- What does an IPv4 address look like?
- Network and subnetwork masks
- Ways to write a network or subnet mask
- Network, node, and broadcast addresses
- Classes of IPv4 addresses
    - IPv4 address classes
    - Network vs. node (host) bits
    - RFC 1918 (private) addresses
    - Local vs. remote addresses
- Classless addresses
- Lessons Learned

Before we can start to subnet IPv4 address spaces, we need to look at what an IPv4 address is used for, how it looks, how to write one down correctly through the use of network masks, and what role address classes formerly served.

What Are IPv4 Addresses Used For?

IPv4 addressing was created to allow one specific digital device to communicate directly with another specific digital device. This is called *unicast* communication. But the designers of IPv4 addressing also wanted a way for one device to communicate with all of the devices within its network at the same time with the same message. If there are ten devices on your network, and one of them wants to send the same message to every other device, using unicast communication would require that the first device send out nine copies of the same message to the other devices. That is a waste of bandwidth. But by using *broadcast* communication, the first device sends out the message once and all nine other hosts receive it. This is a much more efficient use of your bandwidth. Devices designed to use IPv4 addresses are programmed to accept incoming messages to both their unicast and broadcast addresses.

An IPv4 address is known as a *hierarchical* address because it can be divided into multiple, smaller domains, each of which then defines a specific aspect of the device. An IPv4 address, in conjunction with a subnet mask, tells you two pieces of information: the network in which a device is located, and the unique identifier of the device itself.

This is different from a MAC address, which is called a *flat* address; it only tells you about the device itself. A MAC address gives you the unique identifier of the device, but it cannot tell you where in the network this device is located. An IPv4 address is able to tell you both where in the network a device is located and the unique identity of the device itself.

What Does an IPv4 Address Look Like?

The way that a computer or other digital device sees an IPv4 address and the way humans see an IPv4 address are different. A digital device sees an IPv4 address as a 32-bit number. But humans have devised a way to convert this 32-bit number into something easier to look at and work with. For humans, an IPv4 address is a 32-bit number that is broken down in four groups of 8 bits each. Each group of 8 bits is called an octet (or a byte), and the four octets are separated by periods:

11000000.10101000.00000001.00000001

Because we are more comfortable working with the decimal number system, these binary octets are then converted into decimal for ease of use:

192.168.1.1

Network and Subnetwork Masks

Because IPv4 addresses are hierarchical, there needs to be some way of distinguishing between the network portion of the address and the host portion. To accomplish this, an IPv4 address must be paired with a *network mask*, which is also sometimes called a *subnetwork mask* or simply a *subnet mask* or *mask*. A network mask is a 32-bit string of binary characters that defines which portion of the address represents the network and which portion of the address represents the specific host.

The rules for creating a network mask are as follows:

1. A 1 in the subnet mask defines a network bit.

2. A 0 in the subnet mask defines a host bit.

So an IPv4 address by itself really tells us nothing. We can make some guesses based on other information that will come later in this chapter, but for now we need to have a subnet mask paired with an IPv4 address in order for us to know where the device is located and what its unique identity is.

NOTE: Historically, networks were broken down and divided up for use through the use of address classes (which will be discussed later in this chapter). This was accomplished through the use of network masks. When people starting breaking down larger networks into smaller network known as subnetworks, the term *subnetwork mask* came into use. All of these terms are now used interchangeably.

Ways to Write a Network or Subnet Mask

There are three ways to denote a subnet mask:

- In binary: 11111111.11111111.11111111.00000000
- In decimal: 255.255.255.0
- In slash notation: /24

Each of these says the same thing: the first 24 bits of the 32-bit mask are 1, and the last 8 bits of the mask are 0. This tells us that when compared to an IPv4 address, the first 24 bits define the network portion of the address, and the last 8 bits define the host portion. If we look back at the address from the beginning of this chapter and add to it this subnet mask, we get the following:

11000000.10101000.00000001.00000001 (address)

11111111.11111111.11111111.00000000 (subnet mask)

192.168.1.1 255.255.255.0 (address and then mask separated by a single space)

192.168.1.1/24 (address and mask using the / to separate address from mask)

In all three cases we can say that the network portion of the address 192.168.1.1/24 is 192.168.1 and that the host portion is .1.

Network, Node, and Broadcast Addresses

When IPv4 addressing was first designed, there needed to be a way for devices and people to recognize whether an address referred to a specific device or to a group of devices. Some rules were created to help determine the difference between a network address and a node (or host) address:

In an IP address:

- A host portion of all binary 0s refers to the network itself.
- A host portion with a combination of binary 0s and 1s refers to a specific host.
- A host portion of all binary 1s refers to the *broadcast address* of the network.

So what is the difference between these three?

- A *network address* defines the entire network and all of the hosts inside it. This address cannot be assigned to a specific device.
- A *host address* defines one specific device inside of that network. This address can be assigned to a single device.
- A *broadcast address* represents all of the hosts within a specific network. All devices within the network are programmed to accept messages sent to this address.

Going back to our original example of 192.168.1.1/24, what information can we determine now about this address? We already know that based on the mask of /24, the network portion is 192.168.1 and the host portion is .1. We know that the last 8 bits are host bits. Knowing what we now know about network, host, and broadcast addresses:

1. All 0s in the host portion of the address is 00000000 and that equals 0 in decimal, so the network address of this specific network is 192.168.1.0.

2. The host portion in binary is 00000001, which equals 1 in decimal, so this host is the first host in the network.

3. The range of hosts in this network runs from 00000001 to 11111110 in binary, or from 1 to 254 in decimal. There are 254 unique addresses in this network for devices.

4. The broadcast address of this network is 11111111, or all 1s in the host portion. This means that the broadcast address for this network is .255.

In chart form this network would look like:

Network Address	Range of Valid Hosts	Broadcast Address
192.168.1.0	192.168.1.1–192.168.1.254	192.168.1.255

Classes of IPv4 Addresses

IPv4 addresses were originally divided into five different classes according to size. These classes are no longer officially used because concepts such as classless interdomain routing (CIDR) and the mere fact that no more addresses are left to hand out have made address classes a moot point. But the terminology still remains out there and many IT professionals learned using this system, so it is a good starting point for understanding networks and, ultimately, subnetting of networks.

Address classes were broken down based on a concept called the *leading bit pattern* of the first octet of an IPv4 address. Remember that a machine reads IPv4 addressing as a single 32-bit number, so the patterns that were developed were all based in binary. This makes for some not-so-obvious decimal groupings of addresses.

Classes were named A through E and had the characteristics described in Table 1-1.

TABLE 1-1 IPv4 Address Classes

Class	Leading Bit Pattern (First Octet) (in Binary)	First Octet (in Decimal)	Notes
A	0xxxxxxx (x refers to the remaining bits in the octet and can be either 0 or 1)	0–127	0 is invalid 10.0.0.0/8 is reserved for private, internal routing only (RFC 1918) 127 is reserved for loopback testing

Class	Leading Bit Pattern (First Octet) (in Binary)	First Octet (in Decimal)	Notes
B	10xxxxxx	128–191	172.16.0.0/12 is reserved for private, internal routing only (RFC 1918)
C	110xxxxx	192–223	192.168.0.0/16 is reserved for private, internal routing only (RFC 1918)
D	1110xxxx	224–239	Reserved for multicasting; cannot be assigned to unicast hosts
E	1111xxxx	240–255	Reserved for future use/testing

Classes A, B, and C are the only classes that can be used for unicast communication. Class D is used for *multicast communication*, which means that one device can communicate with a specific group of hosts within a network (unlike broadcast communication, which is one device communicating with all hosts within a network). Class E is reserved for future use and/or testing. Class E addresses will never be released for unicast communication.

Network vs. Node (Host) Bits

Within Classes A to C, the four octets of an IPv4 address were broken down to either network octets or node (or host) octets. The following chart shows how the classes were broken down into network bits, called N bits, or node (host) bits, called H bits. The default subnet mask is also shown as well:

Address Class	Octet 1	Octet 2	Octet 3	Octet 4	Default Network Mask
A	NNNNNNNN	HHHHHHHH	HHHHHHHH	HHHHHHHH	/8 or 255.0.0.0
B	NNNNNNNN	NNNNNNNN	HHHHHHHH	HHHHHHHH	/16 or 255.255.0.0
C	NNNNNNNN	NNNNNNNN	NNNNNNNN	HHHHHHHH	/24 or 255.255.255.0

This chart tells us more about the sizing of the different classes of addresses:

- A Class A network has 24 bits that are used for assigning to hosts. 2^{24} = 16,777,216 addresses. Removing two hosts for network identification (all 0s in the host portion) and broadcast communication (all 1s in the host portion), that means that every class A network can host 16,777,214 unique devices in a single network.

- A Class B network has 16 bits available for host assignment. $2^{16} = 65,536$ hosts. Subtract the two addresses reserved for network and broadcast and you have 65,534 valid hosts per Class B network.

- A Class C network has 8 bits available for hosts. $2^8 = 256$ hosts. Subtracting the two addresses reserved for network and broadcast leaves you with 254 valid hosts per Class C network.

Combining this knowledge with the information from the IPv4 address class chart, we can make the following conclusions as well:

- There are 126 valid Class A networks of 16,777,214 valid hosts each.

- There are 16,384 valid Class B networks of 65,534 valid hosts each.

  - We get 16,384 from the fact that the first 16 bits of the address are network bits, but the first two of them are fixed in the pattern of 10. This means that we have 14 bits left of valid network bits ranging from 10**000000.00000000** (which is 128.0) to 10**111111.11111111** (which is 191.255). That is 16,384 different networks.

- There are 2,097,152 Class C networks of 254 valid hosts each.

  - Again, we take the chart that says there are 24 N bits (network bits) and three of those bits are fixed to the pattern of 110. That means we have 21 bits left to assign to networks in the range of 110**00000.00000000.00000000** (192.0.0) to 110**11111.11111111.11111111** (223.255.255). That is 2,097,152 individual networks.

Looking back at our original example in this chapter of 192.168.1.1/24, we now know more information:

- The network is a Class C network.

- The network is in an RFC 1918 network, which means that it is private, and can only be routed internally within the network.

RFC (Private) 1918 Addresses

RFC 1918 addresses were created to help slow down the depletion of IPv4 addresses. Addresses that are part of RFC 1918 are to be used on private, internal networks only. They can be routed within the network, but they are not allowed out onto the public Internet. Most companies and homes today use private, internal RFC 1918 addresses in their networks. In order for a device that is using an RFC 1918 address to get out onto the public Internet, that address has to go through a device that uses Network Address Translation (NAT) and have its private address translated into an acceptable public address.

NOTE: When originally designed, the use of 32-bit numbers was chosen because it was felt that 2^{32} number of unique addresses was so large that it would never be reached. This is why an entire Class A network (127) was reserved for loopback testing. You only need one address to test your loopback, but over 16 *million* addresses were reserved for this test—24 host bits means 2^{24} addresses were reserved when only one was needed. 16,777,216 addresses may seem like a lot, but remember that two of those addresses are reserved for the network and broadcast addresses, so you only have 16,777,214 addresses reserved for loopback testing. Feel better?

NOTE: The original design of address classes had a network of all binary 0s or all binary 1s as invalid. Early devices used a string of 0s or 1s as internal communication codes so they could not be used. Therefore, network 0.0.0.0 and network 255.x.x.x are invalid. The 255 network is part of the reserved Class E network space, so we never had a chance to use it, but the 0 network is part of the assignable Class A address space. Another 16,777,214 addresses lost.

NOTE: We originally thought that 2^{32} bit addressing space would be impossible to reach, but the advent of the public Internet and then mobile devices led to the need for a new addressing scheme. IPv6 is a 128-bit-wide addressing space. Reaching the limit of 2^{128} addresses might seem impossible, but with the Internet of Things (IoT) becoming a reality, is it impossible?

Local vs. Remote Addresses

When two addresses are in the same network, they are said to be *local* to each other. When two addresses are in different networks, they are said to be *remote* to each other. Why is this distinction important? Local devices can communicate directly with each other. Remote devices will need a Layer 3 device (such as a router or a Layer 3 switch) to facilitate communication between the two endpoints. A common mistake that occurs in networks once they are subnetted down to smaller networks is that devices that were once local to each other are now remote, and without that Layer 3 device to assist, communication no longer occurs. Being able to tell if two devices are local or remote is a valuable tool to assist in troubleshooting communication problems.

Classless Addressing

Although classful addressing was originally used in the early days of IPv4 usage, it was quickly discovered that there were inefficiencies in this rigid scheme—who really needs 16 million hosts in a single network? Concepts such as CIDR and variable-length subnet masking (VLSM) were created to allow for a more efficient distribution of IPv4 addresses and networks.

In classless addressing, the rules all stay the same except for one: the size of the default network mask. With classless addressing, the network mask can be changed from the default sizes of /8, /16, or /24 in order to accommodate whatever size of network is required. For example, a common practice is to take a Class A network, such as the RFC 1918 10.0.0.0 network, and break that one large network into smaller, more manageable networks. So instead of a single network of 16.7 million hosts, we can use a /16 mask and create 256 networks of 65,534 hosts each:

10.0.0.0/8 (8 N bits 24 H bits)	=	One network of 16,777, 214 hosts (2^{24} bits for hosts)	10.0.0.1–10.255.255.254
10.0.0.0/16 (16 N bits and 16 H bits)	=	One network of 65,534 hosts (2^{16} bits for hosts)	10.0.0.1–10.0.255.254
10.1.0.0/16	=	One network of 65,534 hosts	10.1.0.1–10.1.255.254
10.2.0.0/16	=	One network of 65,534 hosts	10.2.0.1–10.2.255.254
...			
10.254.0.0/16	=	One network of 65,534 hosts	10.254.0.1–10.254.255.254
10.255.0.0/16	=	One network of 65,534 hosts	10.255.0.1–10.255.255.255

We can prove this using binary to show that no other rules of IP addressing have changed:

		N/H Bits	N Bits	H Bits
10.0.0.0/8	=	8 N and 24 H bits	00001010	00000000 00000000 00000000
10.0.0.0/16	=	16 N and 16 H bits	00001010 00000000	00000000 00000000
10.1.0.0/16	=	16 N and 16 H bits	00001010 00000001	00000000 00000000
10.2.0.0/16	=	16 N and 16 H bits	00001010 00000010	00000000 00000000
...				
10.254.0.0/16	=	16 N and 16 H bits	00001010 11111110	00000000 00000000
10.255.0.0/16	=	16 N and 16 H bits	00001010 11111111	00000000 00000000

The rules for network, valid host, and broadcast addresses have also not changed. If we take the column of H bits above and break it down further we get the following:

Address	Host Bit Portion of Address		
	Network Address (All 0s in H bits)	Range of Valid Hosts	Broadcast Address (All 1s in H bits)
10.0.0.0/16	00000000 00000000 (10.0.*0.0*)	00000000 00000001– 11111111 11111110 (10.0.*0.1*–10.0.255.254)	11111111 11111111 (10.0.*255.255*)
10.1.0.0/16	00000000 00000000 (10.1.*0.0*)	00000000 00000001– 11111111 11111110 (10.1.*0.1*–10.1.255.254)	11111111 11111111 (10.1.*255.255*)

Another common occurrence is to take a Class A network and use the Class C default mask to create 65,536 networks of 254 hosts per network:

10.0.0.0/8 (8 N bits and 24 H bits)	=	One network of 16,777, 214 hosts (2^{24} bits for hosts)	10.0.0.1–10.255.255.254
10.0.0.0/24 (16 N and 16 H bits)	=	One network of 254 hosts (2^{16} bits for hosts)	10.0.0.1–10.0.0.254
10.0.1.0/24	=	One network of 254 hosts	10.0.1.1–10.0.1.254
10.0.2.0/24	=	One network of 254 hosts	10.0.2.1–10.0.2.254
...			
10.255.254.0/24	=	One network of 254 hosts	10.255.254.1–10.255.255.254
10.255.255.0/24	=	One network of 254 hosts	10.255.255.1–10.255.255.254

A third common occurrence is to take a Class B network and use the Class C default Mask to create 256 networks of 254 hosts per network:

172.16.0.0/16 (16 N bits and 16 H bits)	=	One network of 65,534 hosts (216 bits for hosts)	10.0.0.1–10.255.255.254
172.16.0.0/24 (16 N and 16 H bits)	=	One network of 254 hosts (28 bits for hosts)	172.16.0.1–172.16.0.254
172.16.1.0/24	=	One network of 254 hosts	172.16.1.1–172.16.1.254
172.16.2.0/24	=	One network of 254 hosts	172.16.2.1–172.16.2.254
...			
172.16.254.0/24	=	One network of 254 hosts	172.16.254.1–172.16.255.254
172.16.255.0/24	=	One network of 254 hosts	172.16.255.1–172.16.255.254

What we have done in all of these examples is to break down one large network into many smaller, more manageable networks. This is known as *subnetting*. The next section will show you how to do this for any size of subnet that you may require.

Lessons Learned

You've learned that a simple IPv4 address and network mask tells us a lot of information.

192.168.1.1/24 tells us:

- The network portion of this address is 192.168.1.
  - We denote this as 192.168.1.0/24.
- The host portion of this address is .1.
- There are 254 valid hosts on this network ranging from .1 to .254.
  - This is the first valid host in this network.
  - Hosts with addresses of .2 to .254 in this network would be local to this device and therefore able to communicate directly with .1.

- The broadcast address for this network is 192.168.1.255/24.
    - All 1s in the host portion is 11111111 or .255.
- This is a Class C address.
    - First octet has a number of 192.
- This is a valid address that can be assigned to a device for unicast communication.
    - It is not reserved for other use like a Class D multicasting address.
  - It is not reserved for testing/future use like a Class E address.
  - It is not reserved for loopback testing.
  - It is not invalid like the 0 or 255 networks.
    - It is not a network address or a broadcast address.

CHAPTER 2

How to Subnet IPv4 Addresses

This chapter provides information concerning the following topics:

- Subnetting a Class C network using binary
- Subnetting a Class B network using binary
- Binary ANDing
    - So Why AND?
    - Shortcuts in binary ANDing

In the previous chapter, we looked at how IPv4 addressing works, and the idea that it is possible to break a single large networks into multiple smaller networks for more flexibility in your network design. This chapter shows you how to perform this task. This is known as *subnetting*.

NOTE: Some students (and working IT professionals) are intimidated by subnetting because it deals with math; more specifically, binary math. While some people pick this up quickly, some take more time than others. And this is OK. Just keep practicing. The ability to subnet IPv4 addresses is a key skill that is required to pass the CCNA 200-301 exam. This makes some people nervous during an exam. Just remember that this is math, and therefore there has to be an absolute correct answer. If you follow the steps, you will come up with the correct answer. I always tell my students that subnetting and working with binary should be the easiest questions you have on an exam, because you know that if you follow the steps you will arrive at the correct answer. Keep calm, remember the rules, and you will be fine. After all, it's just math, and math is easy.

NOTE: Remember from the previous chapter that there are network bits (N bits) and host bits (H bits) in an IPv4 address and they follow a specific pattern:

Octet #	1	2	3	4
Class A Address	N	H	H	H
Class B Address	N	N	H	H
Class C Address	N	N	N	H

All 0s in host portion = network or subnetwork address

All 1s in host portion = broadcast address

Combination of 1s and 0s in host portion = valid host address

To subnet a network address space, we will use the following formulae:

2^N (where N is equal to the number of network bits borrowed)	Number of *total* subnets created
2^H (where H is equal to the number of host bits)	Number of *total* hosts per subnet
$2^H - 2$	Number of *valid* hosts per subnet

Subnetting a Class C Network Using Binary

You have an address of 192.168.100.0 /24. You need nine subnets. What is the IP plan of network numbers, broadcast numbers, and valid host numbers? What is the subnet mask needed for this plan?

You cannot use N bits, only H bits. Therefore, ignore 192.168.100. These numbers cannot change. You only work with host bits. You need to borrow some host bits and turn them into network bits (or in this case, subnetwork bits; I use the variable N to refer to both network and subnetwork bits).

Step 1.   Determine how many H bits you need to borrow to create nine valid subnets.

$2^N \geq 9$

N = 4, so you need to borrow 4 H bits and turn them into N bits.

Start with 8 H bits	HHHHHHHH
Borrow 4 bits	NNNNHHHH

Step 2.   Determine the first subnet in binary.

0000HHHH	
00000000	All 0s in host portion = subnetwork number
00000001	First valid host number
00000010	Second valid host number
00000011	Third valid host number
...	
00001110	Last valid host number
00001111	All 1s in host portion = broadcast number

Step 3.   Convert binary to decimal.

00000000 = 0	Subnetwork number
00000001 = 1	First valid host number
00000010 = 2	Second valid host number
00000011 = 3	Third valid host number
. ...	
00001110 = 14	Last valid host number
00001111 = 15	All 1s in host portion = broadcast number

Step 4.   Determine the second subnet in binary.

0001HHHH	
0001**0000**	All 0s in host portion = subnetwork number
0001**0001**	First valid host number
0001**0010**	Second valid host number
...	
0001**1110**	Last valid host number
0001**1111**	All 1s in host portion = broadcast number

Step 5.   Convert binary to decimal.

00010000 = 16	Subnetwork number
00010001 = 17	First valid host number
...	
00011110 = 30	Last valid host number
00011111 = 31	All 1s in host portion = broadcast number

Step 6.   Create an IP plan table.

Subnet	Network Number	Range of Valid Hosts	Broadcast Number
1	0	1–14	15
2	16	17–30	31
3	32	33–46	47

Notice a pattern? Counting by 16.

Step 7.   Verify the pattern in binary. (The third subnet in binary is used here.)

0010HHHH	Third subnet
00100000 = **32**	Subnetwork number
00100001 = **33**	First valid host number
00100010 = **34**	Second valid host number
...	
00101110 = **46**	Last valid host number
00101111 = **47**	Broadcast number

Step 8.   Finish the IP plan table.

Subnet	Network Address (0000)	Range of Valid Hosts (0001–1110)	Broadcast Address (1111)
1 (0000)	192.168.100.**0**	192.168.100.**1**– 192.168.100.**14**	192.168.100.**15**
2 (0001)	192.168.100.**16**	192.168.100.**17**– 192.168.100.**30**	192.168.100.**31**

Subnet	Network Address (0000)	Range of Valid Hosts (0001–1110)	Broadcast Address (1111)
3 (0010)	192.168.100.**32**	192.168.100.**33**–192.168.100.**46**	192.168.100.**47**
4 (0011)	192.168.100.**48**	192.168.100.**49**–192.168.100.**62**	192.168.100.**63**
5 (0100)	192.168.100.**64**	192.168.100.**65**–192.168.100.**78**	192.168.100.**79**
6 (0101)	192.168.100.**80**	192.168.100.**81**–192.168.100.**94**	192.168.100.**95**
7 (0110)	192.168.100.**96**	192.168.100.**97**–192.168.100.**110**	192.168.100.**111**
8 (0111)	192.168.100.**112**	192.168.100.**113**–192.168.100.**126**	192.168.100.**127**
9 (1000)	192.168.100.**128**	192.168.100.**129**–192.168.100.**142**	192.168.100.**143**
10 (1001)	192.168.100.**144**	192.168.100.**145**–192.168.100.**158**	192.168.100.**159**
11 (1010)	192.168.100.**160**	192.168.100.**161**–192.168.100.**174**	192.168.100.**175**
12 (1011)	192.168.100.**176**	192.168.100.**177**–192.168.100.**190**	192.168.100.**191**
13 (1100)	192.168.100.**192**	192.168.100.**193**–192.168.100.**206**	192.168.100.**207**
14 (1101)	192.168.100.**208**	192.168.100.**209**–192.168.100.**222**	192.168.100.**223**
15 (1110)	192.168.100.**224**	192.168.100.**225**–192.168.100.**238**	192.168.100.**239**
16 (1111)	192.168.100.**240**	192.168.100.**241**–192.168.100.**254**	192.168.100.**255**
Quick Check	Always an even number	First valid host is always an odd # Last valid host is always an even #	Always an odd number

Use any nine subnets—the rest are for future growth.

Step 9.   Calculate the subnet mask. The default subnet mask for a Class C network is as follows:

Decimal	Binary
255.255.255.0	11111111.11111111.11111111.00000000

1 = Network or subnetwork bit

0 = Host bit

You borrowed 4 bits; therefore, the new subnet mask is the following:

11111111.11111111.11111111.**1111**0000	255.255.255.**240**

NOTE: You subnet a Class B network or a Class A network using exactly the same steps as for a Class C network; the only difference is that you start with more H bits.

Subnetting a Class B Network Using Binary

You have an address of 172.16.0.0 /16. You need nine subnets. What is the IP plan of network numbers, broadcast numbers, and valid host numbers? What is the subnet mask needed for this plan?

You cannot use N bits, only H bits. Therefore, ignore 172.16. These numbers cannot change.

Step 1. Determine how many H bits you need to borrow to create nine valid subnets.

$2^N \geq 9$

N = 4, so you need to borrow 4 H bits and turn them into N bits.

Start with 16 H bits	HHHHHHHHHHHHHHHH (Remove the decimal point for now)
Borrow 4 bits	**NNNN**HHHHHHHHHHHH

Step 2. Determine the first valid subnet in binary (without using decimal points).

0000HHHHHHHHHHHH	
0000000000000000	Subnet number
0000000000000001	First valid host
...	
0000111111111110	Last valid host
0000111111111111	Broadcast number

Step 3. Convert binary to decimal (replacing the decimal point in the binary numbers).

00000000.00000000 = 0.0	Subnetwork number
00000000.00000001 = 0.1	First valid host number
...	
00001111.11111110 = 15.254	Last valid host number
00001111.11111111 = 15.255	Broadcast number

Step 4. Determine the second subnet in binary (without using decimal points).

0001HHHHHHHHHHHH	
0001000000000000	Subnet number
0001000000000001	First valid host
...	
0001111111111110	Last valid host
0001111111111111	Broadcast number

Step 5.   Convert binary to decimal (returning the decimal point in the binary numbers).

00010000.00000000 = 16.0	Subnetwork number
00010000.00000001 = 16.1	First valid host number
...	
00011111.11111110 = 31.254	Last valid host number
00011111.11111111 = 31.255	Broadcast number

Step 6.   Create an IP plan table.

Subnet	Network Number	Range of Valid Hosts	Broadcast Number
1	0.0	0.1–15.254	15.255
2	16.0	16.1–31.254	31.255
3	32.0	32.1–47.254	47.255

Notice a pattern? Counting by 16.

Step 7.   Verify the pattern in binary. (The third subnet in binary is used here.)

0010**HHHHHHHHHHHH**	Third valid subnet
0010**0000.00000000 = 32.0**	Subnetwork number
0010**0000.00000001 = 32.1**	First valid host number
...	
0010**1111.11111110 = 47.254**	Last valid host number
0010**1111.11111111 = 47.255**	Broadcast number

Step 8.   Finish the IP plan table.

Subnet	Network Address (0000)	Range of Valid Hosts (0001–1110)	Broadcast Address (1111)
1 (0000)	172.16.**0.0**	172.16.**0.1**–172.16.**15.254**	172.16.**15.255**
2 (0001)	172.16.**16.0**	172.16.**16.1**–172.16.**31.254**	172.16.**31.255**
3 (0010)	172.16.**32.0**	172.16.**32.1**–172.16.**47.254**	172.16.**47.255**
4 (0011)	172.16.**48.0**	172.16.**48.1**–172.16.**63.254**	172.16.**63.255**
5 (0100)	172.16.**64.0**	172.16.**64.1**–172.16.**79.254**	172.16.**79.255**
6 (0101)	172.16.**80.0**	172.16.**80.1**–172.16.**95.254**	172.16.**95.255**
7 (0110)	172.16.**96.0**	172.16.**96.1**–172.16.**111.254**	172.16.**111.255**

Subnet	Network Address (0000)	Range of Valid Hosts (0001–1110)	Broadcast Address (1111)
8 (0111)	172.16.**112.0**	172.16.**112.1**– 172.16.**127.254**	172.16.**127.255**
9 (1000)	172.16.**128.0**	172.16.**128.1**– 172.16.**143.254**	172.16.**143.255**
10 (1001)	172.16.**144.0**	172.16.**144.1**– 172.16.**159.254**	172.16.**159.255**
11 (1010)	172.16.**160.0**	172.16.**160.1**– 172.16.**175.254**	172.16.**175.255**
12 (1011)	172.16.**176.0**	172.16.**176.1**– 172.16.**191.254**	172.16.**191.255**
13 (1100)	172.16.**192.0**	172.16.**192.1**– 172.16.**207.254**	172.16.**207.255**
14 (1101)	172.16.**208.0**	172.16.**208.1**– 172.16.**223.254**	172.16.**223.255**
15 (1110)	172.16.**224.0**	172.16.**224.1**– 172.16.**239.254**	172.16.**239.255**
16 (1111)	172.16.**240.0**	172.16.**240.1**– 172.16.**255.254**	172.16.**255.255**
Quick Check	**Always in form even #.0**	**First valid host is always even #.1** **Last valid host is always odd #.254**	**Always odd #.255**

Use any nine subnets—the rest are for future growth.

Step 9.   Calculate the subnet mask. The default subnet mask for a Class B network is as follows:

Decimal	Binary
255.255.0.0	11111111.11111111.00000000.00000000

1 = Network or subnetwork bit

0 = Host bit

You borrowed 4 bits; therefore, the new subnet mask is the following:

11111111.11111111.**1111**0000.00000000	255.255.**240**.0

Binary ANDing

Binary ANDing is the process of performing multiplication to two binary numbers. In the decimal numbering system, ANDing is addition: 2 and 3 equals 5. In decimal, there are an infinite number of answers when ANDing two numbers together. However, in the binary numbering system, the AND function yields only two possible outcomes, based on four different combinations. These outcomes, or answers, can be displayed in what is known as a truth table:

0 and 0 = 0

1 and 0 = 0

0 and 1 = 0

1 and 1 = 1

You use ANDing most often when comparing an IP address to its subnet mask. The end result of ANDing these two numbers together is to yield the network number of that address.

Question 1

What is the network number of the IP address 192.168.100.115 if it has a subnet mask of 255.255.255.240?

Answer

Step 1.   Convert both the IP address and the subnet mask to binary:

```
192.168.100.115 = 11000000.10101000.01100100.01110011
255.255.255.240 = 11111111.11111111.11111111.11110000
```

Step 2.   Perform the AND operation to each pair of bits—1 bit from the address ANDed to the corresponding bit in the subnet mask. Refer to the truth table for the possible outcomes.

```
192.168.100.115 = 11000000.10101000.01100100.01110011
255.255.255.240 = 11111111.11111111.11111111.11110000
ANDed result    = 11000000.10101000.01100100.01110000
```

Step 3.   Convert the answer back into decimal:

```
11000000.10101000.01100100.01110000 = 192.168.100.112
The IP address 192.168.100.115 belongs to the 192.168.100.112
network when a mask of 255.255.255.240 is used.
```

Question 2

What is the network number of the IP address 192.168.100.115 if it has a subnet mask of 255.255.255.192?

(Notice that the IP address is the same as in Question 1, but the subnet mask is different. What answer do you think you will get? The same one? Let's find out!)

Answer

Step 1.   Convert both the IP address and the subnet mask to binary:

```
192.168.100.115 = 11000000.10101000.01100100.01110011
255.255.255.192 = 11111111.11111111.11111111.11000000
```

Step 2. Perform the AND operation to each pair of bits—1 bit from the address ANDed to the corresponding bit in the subnet mask. Refer to the truth table for the possible outcomes.

```
192.168.100.115 = 11000000.10101000.01100100.01110011
255.255.255.192 = 11111111.11111111.11111111.11000000
ANDed result    = 11000000.10101000.01100100.01000000
```

Step 3. Convert the answer back into decimal:

```
11000000.10101000.01100100.01110000 = 192.168.100.64
```

The IP address 192.168.100.115 belongs to the 192.168.100.64 network when a mask of 255.255.255.192 is used.

So Why AND?

Good question. The best answer is to save you time when working with IP addressing and subnetting. If you are given an IP address and its subnet, you can quickly find out what subnetwork the address belongs to. From here, you can determine what other addresses belong to the same subnet. Remember that if two addresses are in the same network or subnetwork, they are considered to be *local* to each other and can therefore communicate directly with each other. Addresses that are not in the same network or subnetwork are considered to be *remote* to each other and must therefore have a Layer 3 device (like a router or Layer 3 switch) between them to communicate.

Question 3

What is the broadcast address of the IP address 192.168.100.164 if it has a subnet mask of 255.255.255.248?

Answer

Step 1. Convert both the IP address and the subnet mask to binary:

```
192.168.100.164 = 11000000.10101000.01100100.10100100
255.255.255.248 = 11111111.11111111.11111111.11111000
```

Step 2. Perform the AND operation to each pair of bits—1 bit from the address ANDed to the corresponding bit in the subnet mask. Refer to the truth table for the possible outcomes:

```
192.168.100.164 = 11000000.10101000.01100100.10100100
255.255.255.248 = 11111111.11111111.11111111.11111000
ANDed result    = 11000000.10101000.01100100.10100000
                = 192.168.100.160 (Subnetwork #)
```

Step 3. Separate the network bits from the host bits:

```
255.255.255.248 = /29. The first 29 bits are network/
subnetwork bits; therefore,
```

11000000.10101000.01100100.10100000. The last 3 bits are host bits.

Step 4.   Change all host bits to 1. Remember that all 1s in the host portion are the broadcast number for that subnetwork.

```
11000000.10101000.01100100.10100111
```

Step 5.   Convert this number to decimal to reveal your answer:

```
11000000.10101000.01100100.10100111 = 192.168.100.167
```

```
The broadcast address of 192.168.100.164 is 192.168.100.167
when the subnet mask is 255.255.255.248.
```

Shortcuts in Binary ANDing

Remember that I said ANDing is supposed to save you time when working with IP addressing and subnetting? Well, there are shortcuts when you AND two numbers together:

- An octet of all 1s in the subnet mask results in the answer being the same octet as in the IP address.

- An octet of all 0s in the subnet mask results in the answer being all 0s in that octet.

Question 4

To what network does 172.16.100.45 belong, if its subnet mask is 255.255.255.0?

Answer

172.16.100.0

Proof

Step 1.   Convert both the IP address and the subnet mask to binary:

```
172.16.100.45  = 10101100.00010000.01100100.00101101
255.255.255.0  = 11111111.11111111.11111111.00000000
```

Step 2.   Perform the AND operation to each pair of bits—1 bit from the address ANDed to the corresponding bit in the subnet mask. Refer to the truth table for the possible outcomes.

```
172.16.100.45  = 10101100.00010000.01100100.00101101
255.255.255.0  = 11111111.11111111.11111111.00000000
                 10101100.00010000.01100100.00000000
               = 172.16.100.0
```

Notice that the first three octets have the same pattern both before and after they were ANDed. Therefore, any octet ANDed to a subnet mask pattern of 255 is itself! Notice that the last octet is all 0s after ANDing. But according to the truth table, anything ANDed to a 0 is a 0. Therefore, any octet ANDed to a subnet mask pattern of 0 is 0! You should only have to convert those parts of an IP address and subnet mask to binary if the mask is not 255 or 0.

Question 5

To what network does 68.43.100.18 belong if its subnet mask is 255.255.255.0?

Answer

68.43.100.0 (There is no need to convert here. The mask is either 255s or 0s.)

Question 6

To what network does 131.186.227.43 belong if its subnet mask is 255.255.240.0?

Answer

Based on the two shortcut rules, the answer should be

    131.186.???.0

So now you only need to convert one octet to binary for the ANDing process:

    227   = 11100011

    240   = 11110000

    11100000 = 224

Therefore, the answer is 131.186.224.0.

CHAPTER 3

Variable Length Subnet Masking (VLSM)

This chapter provides information concerning the following topics:

- IP subnet zero
- VLSM example

Variable-length subnet masking (VLSM) is the more realistic way of subnetting a network to make the most efficient use of all of the bits.

Remember that when you perform classful (or what I sometimes call classical) subnetting, all subnets have the same number of hosts because they all use the same subnet mask. This leads to inefficiencies. For example, if you borrow 4 bits on a Class C network, you end up with 16 valid subnets of 14 valid hosts per subnet. A point-to-point link to another router only needs 2 hosts, but with classical subnetting, you end up wasting 12 of those hosts. Even with the ability to use NAT and private addresses, where you should never run out of addresses in a network design, you still want to ensure that the IP plan you create is as efficient as possible. This is where VLSM comes into play.

VLSM is the process of "subnetting a subnet" and using different subnet masks for different networks in your IP plan. What you have to remember is that you need to make sure that there is no overlap in any of the addresses.

IP Subnet Zero

Historically, it was always recommended that a subnet of all 0s or a subnet of all 1s not be used. Therefore, the formula of $2^N - 2$ was used to calculate the number of valid subnets created. However, Cisco devices can use those subnets, as long as the command **ip subnet-zero** is in the configuration. This command is on by default in Cisco IOS Software Release 12.0 and later; if it was turned off for some reason, however, you can reenable it by using the following command:

```
Router(config)# ip subnet-zero
```

Now you can use the formula 2^N rather than $2^N - 2$.

2^N	Number of total subnets created	
~~$2^N - 2$~~	~~Number of valid subnets created~~	No longer needed because you have the **ip subnet-zero** command enabled
2^H	Number of total hosts per subnet	
$2^H - 2$	Number of valid hosts per subnet	

NOTE: All of this is explained in great detail in RFC 950, *Internet Standard for Subnetting Procedure* (August 1985).

NOTE: RFC 1878, *Variable Length Subnet Table for IPv4* (December 1995), states, "This practice [of excluding all-zeros and all-ones subnets] is obsolete. Modern software will be able to utilize all definable networks."

VLSM Example

NOTE: Throughout this book, I use serial links between routers to help differentiate these networks from Ethernet networks where hosts usually reside. In today's more modern networks, Ethernet links are used almost exclusively between routers, and serial links are rapidly becoming obsolete. However, using serial links is a very cost-effective way to set up a testing lab for learning purposes; I have seen home labs, school labs, and corporate training labs use serial links for this reason. Even though a simple Ethernet link (10 Mbps) is faster than a serial link (1.544 Mbps), I use serial links in all of my Portable Command Guides to show the difference between an Ethernet link and a point-to-point link to another router.

You follow the same steps in performing VLSM as you did when performing classical subnetting.

Consider Figure 3-1 as you work through an example.

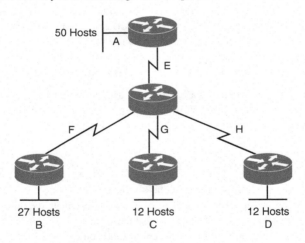

Figure 3-1   Sample Network Needing a VLSM Address Plan

A Class C network—192.168.100.0/24—is assigned. You need to create an IP plan for this network using VLSM.

Once again, you cannot use the N bits—192.168.100. You can use only the H bits. Therefore, ignore the N bits because they cannot change!

The steps to create an IP plan using VLSM for the network illustrated in Figure 3-1 are as follows:

Step 1.   Determine how many H bits will be needed to satisfy the *largest* network.

Step 2.   Pick a subnet for the largest network to use.

Step 3.   Pick the next largest network to work with.

Step 4.   Pick the third largest network to work with.

Step 5.   Determine network numbers for serial links.

The remainder of the chapter details what is involved with each step of the process.

Step 1: Determine How Many H Bits Will Be Needed to Satisfy the *Largest* Network

Network A is the largest network with 50 hosts. Therefore, you need to know how many H bits will be needed:

If $2^H - 2$ = Number of valid hosts per subnet

Then $2^H - 2 \geq 50$

Therefore H = 6 (6 is the smallest valid value for H)

You need 6 H bits to satisfy the requirements of Network A.

If you need 6 H bits and you started with 8 N bits, you are left with $8 - 6 = 2$ N bits to create subnets:

Started with: NNNNNNNN (these are the 8 bits in the fourth octet)

Now have: NNHHHHHH

All subnetting will now have to start at this reference point to satisfy the requirements of Network A.

Step 2: Pick a Subnet for the Largest Network to Use

You have 2 N bits to work with, leaving you with 2^N or 2^2 or 4 subnets to work with:

NN = 00HHHHHH (The Hs = The 6 H bits you need for Network A)

01HHHHHH

10HHHHHH

11HHHHHH

If you add all 0s to the H bits, you are left with the network numbers for the four subnets:

00**000000** = .0

01**000000** = .64

10**000000** = .128

11**000000** = .192

All of these subnets will have the same subnet mask, just like in classful subnetting.

Two borrowed H bits means a subnet mask of

11111111.11111111.11111111.11000000

or

255.255.255.192

or

/26

The /x notation represents how to show different subnet masks when using VLSM.

/8 means that the first 8 bits of the address are network; the remaining 24 bits are H bits.

/24 means that the first 24 bits are network; the last 8 are host. This is either a traditional default Class C address, a traditional Class A network that has borrowed 16 bits, or even a traditional Class B network that has borrowed 8 bits!

Pick *one* of these subnets to use for Network A. The rest of the networks will have to use the other three subnets.

For purposes of this example, pick the .64 network.

00000000 =	.0	
01000000 =	.64	Network A
10000000 =	.128	
11000000 =	.192	

Step 3: Pick the Next Largest Network to Work With

Network B = 27 hosts

Determine the number of H bits needed for this network:

$2^H - 2 \geq 27$

$H = 5$

You need 5 H bits to satisfy the requirements of Network B.

You started with a pattern of 2 N bits and 6 H bits for Network A. You have to maintain that pattern.

Pick one of the remaining /26 networks to work with Network B.

For the purposes of this example, select the .128/26 network:

10000000

But you need only 5 H bits, not 6. Therefore, you are left with

    10**N00000**

where

    10 represents the original pattern of subnetting.

    N represents the extra bit.

    00000 represents the 5 H bits you need for Network B.

Because you have this extra bit, you can create two smaller subnets from the original subnet:

    10**000000**

    10**100000**

Converted to decimal, these subnets are as follows:

    10**000000** =.128

    10**100000** =.160

You have now subnetted a subnet! This is the basis of VLSM.

Each of these sub-subnets will have a new subnet mask. The original subnet mask of /24 was changed into /26 for Network A. You then take one of these /26 networks and break it into two /27 networks:

    10**000000** and 10**100000** both have 3 N bits and 5 H bits.

The mask now equals:

    11111111.11111111.11111111.11100000

or

    255.255.255.224

or

    /27

Pick one of these new sub-subnets for Network B:

    10**000000** /27 = Network B

Use the remaining sub-subnet for future growth, or you can break it down further if needed.

You want to make sure the addresses are not overlapping with each other. So go back to the original table.

00000000 =	.0/26	
01000000 =	.64/26	Network A
10000000 =	.128/26	
11000000 =	.192/26	

You can now break the .128/26 network into two smaller /27 networks and assign Network B.

00000000 =	.0/26	
01000000 =	.64/26	Network A
10000000 =	.128/26	Cannot use because it has been subnetted
10000000 =	.128/27	Network B
10100000 =	.160/27	
11000000 =	.192/26	

The remaining networks are still available to be assigned to networks or subnetted further for better efficiency.

Step 4: Pick the Third Largest Network to Work With

Networks C and Network D = 12 hosts each

Determine the number of H bits needed for these networks:

$2^H - 2 \geq 12$

$H = 4$

You need 4 H bits to satisfy the requirements of Network C and Network D.

You started with a pattern of 2 N bits and 6 H bits for Network A. You have to maintain that pattern.

You now have a choice as to where to put these networks. You could go to a different /26 network, or you could go to a /27 network and try to fit them into there.

For the purposes of this example, select the other /27 network—.160/27:

10100000 (The 1 in the third bit place is no longer bold because it is part of the N bits.)

But you only need 4 H bits, not 5. Therefore, you are left with

101N0000

where

10 represents the original pattern of subnetting.

N represents the extra bit you have.

00000 represents the 5 H bits you need for Networks C and D.

Because you have this extra bit, you can create two smaller subnets from the original subnet:

101**00000**

101**10000**

Converted to decimal, these subnets are as follows:

101**00000** = .160

101**10000** = .176

These new sub-subnets will now have new subnet masks. Each sub-subnet now has 4 N bits and 4 H bits, so their new masks will be

11111111.11111111.11111111.11110000

or

255.255.255.240

or

/28

Pick one of these new sub-subnets for Network C and one for Network D.

000**00000** =	.0/26	
01**000000** =	.64/26	Network A
10**000000** =	.128/26	Cannot use because it has been subnetted
10**000000** =	.128/27	Network B
101**00000** =	.160/27	Cannot use because it has been subnetted
101**00000**	.160/28	Network C
101**10000**	.176/28	Network D
11**000000** =	.192/26	

You have now used two of the original four subnets to satisfy the requirements of four networks. Now all you need to do is determine the network numbers for the serial links between the routers.

Step 5: Determine Network Numbers for Serial Links

All serial links between routers have the same property in that they only need two addresses in a network—one for each router interface.

Determine the number of H bits needed for these networks:

$$2^H - 2 \geq 2$$

$$H = 2$$

You need 2 H bits to satisfy the requirements of Networks E, F, G, and H.

You have two of the original subnets left to work with.

For the purposes of this example, select the .0/26 network:

00000000

But you need only 2 H bits, not 6. Therefore, you are left with

00NNNN00

where

00 represents the original pattern of subnetting.

NNNN represents the extra bits you have.

00 represents the 2 H bits you need for the serial links.

Because you have 4 **N** bits, you can create 16 sub-subnets from the original subnet:

000000**00** = .0/30

000001**00** = .4/30

000010**00** = .8/30

000011**00** = .12/30

000100**00** = .16/30

...

001110**00** = .56/30

001111**00** = .60/30

You need only four of them. You can hold the rest for future expansion or recombine them for a new, larger subnet:

000100**00** = .16/30

000101**00** = .20/30

00**011000** = .24/30

00**011100** = .32/30

…

00**111000** = .56/30

00**111100** = .60/30

The first four of these can be combined into the following:

00**010000** = .16/28

The rest of the /30 subnets can be combined into two /28 networks:

00**100000** = .32/28

00**110000** = .48/28

Or these two subnets can be combined into one larger /27 network:

00**010000** = .32/27

Going back to the original table, you now have the following:

00**000000** =	.0/26	Cannot use because it has been subnetted
00**000000** =	.0/30	Network E
00**000100** =	.4/30	Network F
00**001000** =	.8/30	Network G
00**001100** =	.12/30	Network H
00**010000** =	.16/28	Future growth
00**100000** =	.32/27	Future growth
01**000000** =	.64/26	Network A
10**000000** =	.128/26	Cannot use because it has been subnetted
10**000000** =	.128/27	Network B
10**100000** =	.160/27	Cannot use because it has been subnetted
10**100000**	.160/28	Network C
10**110000**	.176/28	Network D
11**000000** =	.192/26	Future growth

Looking at the plan, you can see that no number is used twice. You have now created an IP plan for the network and have made the plan as efficient as possible, wasting no addresses in the serial links and leaving room for future growth. This is the power of VLSM!

CHAPTER 4

Route Summarization

This chapter provides information concerning the following topics:

- Example for understanding route summarization
- Route summarization and route flapping
- Requirements for route summarization

Route summarization, or supernetting, is needed to reduce the number of routes that a router advertises to its neighbor. Remember that for every route you advertise, the size of your update grows. It has been said that if there were no route summarization, the Internet backbone would have collapsed from the sheer size of its own routing tables back in 1997!

Routing updates, whether done with a distance-vector protocol or a link-state protocol, grow with the number of routes you need to advertise. In simple terms, a router that needs to advertise ten routes needs ten specific lines in its update packet. The more routes you have to advertise, the bigger the packet. The bigger the packet, the more bandwidth the update takes, reducing the bandwidth available to transfer data. But with route summarization, you can advertise many routes with only one line in an update packet. This reduces the size of the update, allowing you more bandwidth for data transfer.

Also, when a new data flow enters a router, the router must do a lookup in its routing table to determine which interface the traffic must be sent out. The larger the routing tables, the longer this takes, leading to more used router CPU cycles to perform the lookup. Therefore, a second reason for route summarization is that you want to minimize the amount of time and router CPU cycles that are used to route traffic.

> **NOTE:** This example is a very simplified explanation of how routers send updates to each other. For a more in-depth description, I highly recommend you go out and read Jeff Doyle and Jennifer Carroll's book *Routing TCP/IP, Volume I*, Second Edition (Cisco Press, 2005). This book has been around for many years and is considered by most to be the authority on how the different routing protocols work. If you are considering continuing on in your certification path to try and achieve the CCIE, you need to buy Doyle's book—and memorize it; it's that good.

Example for Understanding Route Summarization

Refer to Figure 4-1 to assist you as you go through the following explanation of an example of route summarization.

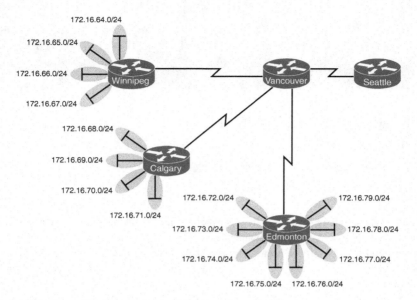

Figure 4-1 Four-City Network Without Route Summarization

As you can see from Figure 4-1, Winnipeg, Calgary, and Edmonton each have to advertise internal networks to the main router located in Vancouver. Without route summarization, Vancouver would have to advertise 16 networks to Seattle. You want to use route summarization to reduce the burden on this upstream router.

Step 1: Summarize Winnipeg's Routes

To do this, you need to look at the routes in binary to see if there are any specific bit patterns that you can use to your advantage. What you are looking for are common bits on the network side of the addresses. Because all of these networks are /24 networks, you want to see which of the first 24 bits are common to all four networks.

172.16.64.0 = **10101100.00010000.01000000**.00000000

172.16.65.0 = **10101100.00010000.01000000**1.00000000

172.16.66.0 = **10101100.00010000.01000001**0.00000000

172.16.67.0 = **10101100.00010000.01000001**1.00000000

Common bits: **10101100.00010000.010000**xx

You see that the first 22 bits of the four networks are common. Therefore, you can summarize the four routes by using a subnet mask that reflects that the first 22 bits are common. This is a /22 mask, or 255.255.252.0. You are left with the summarized address of

172.16.64.0/22

This address, when sent to the upstream Vancouver router, will tell Vancouver: "If you have any packets that are addressed to networks that have the first 22 bits in the pattern of 10101100.00010000.010000xx.xxxxxxxx, then send them to me here in Winnipeg."

By sending one route to Vancouver with this supernetted subnet mask, you have advertised four routes in one line instead of using four lines. Much more efficient!

Step 2: Summarize Calgary's Routes

For Calgary, you do the same thing that you did for Winnipeg—look for common bit patterns in the routes:

>   172.16.68.0 = **10101100.00010000.01000100**.00000000
>
>   172.16.69.0 = **10101100.00010000.01000101**.00000000
>
>   172.16.70.0 = **10101100.00010000.01000110**.00000000
>
>   172.16.71.0 = **10101100.00010000.01000111**.00000000
>
>   Common bits: **10101100.00010000.010001**xx

Once again, the first 22 bits are common. The summarized route is therefore

>   172.16.68.0/22

Step 3: Summarize Edmonton's Routes

For Edmonton, you do the same thing that you did for Winnipeg and Calgary—look for common bit patterns in the routes:

>   172.16.72.0 = **10101100.00010000.01001000**.00000000
>
>   172.16.73.0 = **10101100.00010000.01001001**.00000000
>
>   172.16.74.0 = **10101100.00010000** 01001010.00000000
>
>   172.16.75.0 = **10101100.00010000** 01001011.00000000
>
>   172.16.76.0 = **10101100.00010000.01001100**.00000000
>
>   172.16.77.0 = **10101100.00010000.01001101**.00000000
>
>   172.16.78.0 = **10101100.00010000.01001110**.00000000
>
>   172.16.79.0 = **10101100.00010000.01001111**.00000000
>
>   Common bits: **10101100.00010000.01001**xxx

For Edmonton, the first 21 bits are common. The summarized route is therefore

>   172.16.72.0/21

Figure 4-2 shows what the network looks like, with Winnipeg, Calgary, and Edmonton sending their summarized routes to Vancouver.

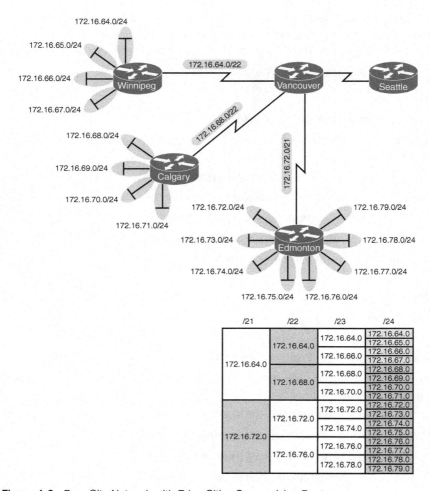

Figure 4-2 Four-City Network with Edge Cities Summarizing Routes

Step 4: Summarize Vancouver's Routes

Yes, you can summarize Vancouver's routes to Seattle. You continue in the same format as before. Take the routes that Winnipeg, Calgary, and Edmonton sent to Vancouver, and look for common bit patterns:

172.16.64.0 = **10101100.00010000.01000000**.00000000

172.16.68.0 = **10101100.00010000.01000**100.00000000

172.16.72.0 = **10101100.00010000.01001**000.00000000

Common bits: **10101100.00010000.0100**xxxx

Because there are 20 bits that are common, you can create one summary route for Vancouver to send to Seattle:

172.16.64.0/20

Vancouver has now told Seattle that in one line of a routing update, 16 different networks are being advertised. This is much more efficient than sending 16 lines in a routing update to be processed.

Figure 4-3 shows what the routing updates would look like with route summarization taking place.

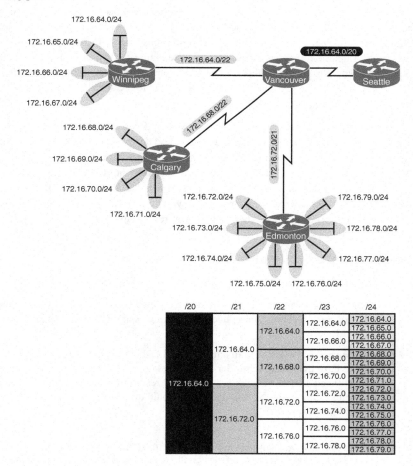

Figure 4-3 Four-City Network with Complete Route Summarization

Route Summarization and Route Flapping

Another positive aspect of route summarization has to do with route flapping. *Route flapping* is when a network, for whatever reason (such as interface hardware failure or misconfiguration), goes up and down on a router, causing that router to constantly advertise changes about that network. Route summarization can help insulate upstream neighbors from these problems.

Consider router Edmonton from Figure 4-1. Suppose that network 172.16.74.0/24 goes down. Without route summarization, Edmonton would advertise Vancouver to remove that network. Vancouver would forward that same message upstream to Calgary, Winnipeg, Seattle, and so on. Now assume the network comes back online a few seconds later. Edmonton would have to send another update informing Vancouver of the change. Each time a change needs to be advertised, the router must use CPU resources. If that route were to flap, the routers would constantly have to update their own tables, as well as advertise changes to their neighbors. In a CPU-intensive protocol such as OSPF, the constant hit on the CPU might make a noticeable change to the speed at which network traffic reaches its destination.

Route summarization enables you to avoid this problem. Even though Edmonton would still have to deal with the route constantly going up and down, no one else would notice. Edmonton advertises a single summarized route, 172.16.72.0/21, to Vancouver. Even though one of the networks is going up and down, this does not invalidate the route to the other networks that were summarized. Edmonton will deal with its own route flap, but Vancouver will be unaware of the problem downstream in Edmonton. Summarization can effectively protect or insulate other routers from route flaps.

Requirements for Route Summarization

To create route summarization, there are some necessary requirements:

- Routers need to be running a classless routing protocol, as they carry subnet mask information with them in routing updates. (Examples are RIP v2, OSPF, EIGRP, IS-IS, and BGP.)

- Addresses need to be assigned in a hierarchical fashion for the summarized address to have the same high-order bits. It does no good if Winnipeg has network 172.16.64.0 and 172.16.67.0 while 172.16.65.0 resides in Calgary and 172.16.66.0 is assigned in Edmonton. No summarization could take place from the edge routers to Vancouver.

TIP: Because most networks use NAT and the RFC 10.0.0.0/8 network internally, it is important when creating your network design that you assign network subnets in a way that they can be easily summarized. A little more planning now can save you a lot of grief later.

CHAPTER 5

IPv6 Addressing—How It Works

This chapter provides information concerning the following topics:

- IPv6: A very brief introduction
- What does an IPv6 address look like?
- Reducing the notation of an IPv6 address
  - Rule 1: Omit leading 0s
  - Rule 2: Omit all-0s hextet
  - Combining rule 1 and rule 2
- Prefix length notation
- IPv6 address types
  - Unicast addresses
    - Global unicast
    - Link-local
    - Loopback
    - Unspecified
    - Unique local
    - IPv4 embedded
  - Multicast addresses
    - Well-known
    - Solicited-node
  - Anycast addresses

NOTE: This chapter is meant to be a very high-level overview of IPv6 addressing. For an excellent overview of IPv6, I strongly recommend you read Rick Graziani's book from Cisco Press: *IPv6 Fundamentals: A Straightforward Approach to Understanding IPv6*, Second Edition. It is a brilliant read, and Rick is an amazing author. I am also very fortunate to call him a friend.

IPv6: A Very Brief Introduction

When IPv4 became a standard in 1980, its 32-bit address field created a theoretical maximum of approximately 4.29 billion addresses (2^{32}). IPv4 was originally conceived as an experiment, and not for a practical implementation, so 4.29 billion was considered to be an inexhaustible amount. But with the growth of the Internet, and the need for individuals and companies to require multiple addresses—your home PC, your cell phone, your tablet, your PC at work/school, your Internet-aware appliances—you can see that something larger than 32-bit address fields would be required. In 1993, the Internet Engineering Task Force (IETF) formed a working group called the IP Next Generation working group. In 1994 the IETF recommended an address size of 128 bits. While many people think that IPv6 is just a way to create more addresses, there are actually many enhancements that make IPv6 a superior choice to IPv4. Again, I recommend Rick Graziani's *IPv6 Fundamentals* as a must-have on your bookshelf for working with IPv6.

What Does an IPv6 Address Look Like?

The way that a computer or other digital device sees an IPv6 address and the way humans see an IPv6 address are different. A digital device looks at an IPv6 address as a 128-bit number. But humans have devised a way to convert this 128-bit number into something easier to look at and work with. For humans, an IPv6 address is a 128-bit number that is written as a string of hexadecimal digits. Hexadecimal is a natural fit for IPv6 addresses because any 4 bits can be represented as a single hexadecimal digit. Two hexadecimal digits represent a single byte, or octet (8 bits). The preferred form of an IPv6 address is $x:x:x:x:x:x:x:x$, where each x is a 16-bit section that can be represented using up to four hexadecimal digits. Each section is separated by a colon (:), as opposed to IPv4 addressing, which uses a period (.) to separate each section. The result is eight 16-bit sections (sometimes called *hextets*) for a total of 128 bits in the address. Figure 5-1 shows this format.

Each 'x' represents up to four hexadecimal digits separated by colons:

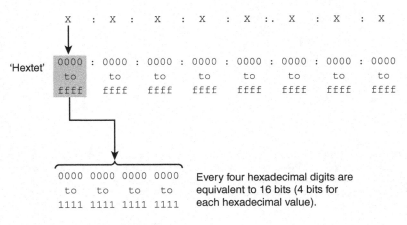

Every four hexadecimal digits are equivalent to 16 bits (4 bits for each hexadecimal value).

Figure 5-1   Format of an IPv6 Address

Showing all the hexadecimal digits in an IPv6 address is the longest representation of the preferred form. The next section shows you two rules for reducing the notation of an IPv6 address in the preferred format for easier use and readability.

TIP: If you need more practice working with hexadecimals and converting between hexadecimal, decimal, and binary, refer to both Appendix A, "How to Count in Decimal, Binary, and Hexadecimal," and Appendix B, "How to Convert Between Number Systems."

Reducing the Notation of an IPv6 Address

Looking at the longest representation of an IPv6 address can be overwhelming:

```
0000:0000:0000:0000:0000:0000:0000:0000
0000:0000:0000:0000:0000:0000:0000:0001
ff02:0000:0000:0000:0000:0000:0000:0001
fe80:0000:0000:0000:a299:9bff:fe18:50d1
2001:0db8:cafe:0001:0000:0000:0000:0200
```

There are two rules for reducing the notation.

Rule 1: Omit Leading 0s

Omit any leading 0s in any hextet (a 16-bit section). This rule applies only to leading 0s and not trailing 0s. Table 5-1 shows examples of omitting leading 0s in a hextet:

TABLE 5-1   Examples of Omitting Leading 0s in a Hextet (Leading 0s in bold; spaces retained)

Format	IPv6 Address
Preferred	**0000**:**0000**:**0000**:**0000**:**0000**:**0000**:**0000**:**0000**
Leading 0s omitted	0: 0: 0: 0: 0: 0: 0: 0 or 0:0:0:0:0:0:0:0
Preferred	**0000**:**0000**:**0000**:**0000**:**0000**:**0000**:**0000**:**000**1
Leading 0s omitted	0: 0: 0: 0: 0: 0: 0: 1 or 0:0:0:0:0:0:0:1
Preferred	ff02:**0000**:**0000**:**0000**:**0000**:**0000**:**0000**:**000**1
Leading 0s omitted	ff02: 0: 0: 0: 0: 0: 0: 1 or ff02:0:0:0:0:0:0:1
Preferred	2001:**0**db8:1111:**000**a:**00**b0:**0000**:9000:**0**200
Leading 0s omitted	2001: db8: 1111: a: b0: 0:9000: 200 or 2001:db8:1111:a:b0:0:9000:200

Rule 2: Omit All-0s Hextet

Use a double colon (::) to represent any single, contiguous string of two or more hextets consisting of all 0s. Table 5-2 shows examples of using the double colon.

TABLE 5-2   Examples of Omitting a Single Contiguous String of All-0s Hextets (0s in Bold Replaced By a Double Colon)

Format	IPv6 Address
Preferred	**0000:0000:0000:0000:0000:0000:0000:0000**
(::) All-0s segments	::
Preferred	**0000:0000:0000:0000:0000:0000:0000:**0001
(::) All-0s segments	::0001
Preferred	ff02:**0000:0000:0000:0000:0000:0000:**0001
(::) All-0s segments	ff02::0001
Preferred	2001:0db8:aaaa:0001:**0000:0000:0000:**0100
(::) All-0s segments	2001:0db8:aaaa:0001::0100
Preferred	2001:0db8:**0000:0000:**abcd:0000:0000:1234
(::) All-0s segments	2001:0db8::abcd:0000:0000:1234

Only a single contiguous string of all 0s can be represented by a double colon; otherwise the address would be ambiguous. Consider the following example:

`2001::abcd::1234`

There are many different possible choices for the preferred address:

```
2001:0000:0000:0000:0000:abcd:0000:1234
2001:0000:0000:0000:abcd:0000:0000:1234
2001:0000:0000:abcd:0000:0000:0000:1234
2001:0000:abcd:0000:0000:0000:0000:1234
```

If two double colons are used, you cannot tell which of these addresses is correct.

If you have an address with more than one contiguous string of 0s, where should you place the double colon? RFC 5952 states that the double colon should represent

- The longest string of all-0s hextets.
- If the strings are of equal value, the first string should use the double colon notation.

Combining Rule 1 and Rule 2

You can combine the two rules to reduce an address even further. Table 5-3 shows examples of this.

TABLE 5-3   Examples of Applying Both Rule 1 and Rule 2 (Leading 0s in bold)

Format	IPv6 Address							
Preferred	**0000**:**0000**:**0000**:**0000**:**0000**:**0000**:**0000**:**0000**							
Leading 0s omitted	0:	0:	0:	0:	0:	0:	0:	0
(::) All-0s segments	::							
Compressed	::							
Preferred	**0000**:**0000**:**0000**:**0000**:**0000**:**0000**:**0000**:**000**1							
Leading 0s omitted	0:	0:	0:	0:	0:	0:	0:	1
(::) All-0s segments	::1							
Compressed	::1							
Preferred	ff02:**0000**:**0000**:**0000**:**0000**:**0000**:**0000**:**000**1							
Leading 0s omitted	ff02:	0:	0:	0:	0:	0:	0:	1
(::) All-0s segments	ff02::1							
Compressed	ff02::1							
Preferred	fe80:**0000**:**0000**:**0000**:a299:9bff:fe18:50d1							
Leading 0s omitted	fe80:	0:	0:	0:a299:9bff:fe18:50d1				
(::) All-0s segments	fe80::a299:9bff:fe18:50d1							
Compressed	fe80::a299:9bff:fe18:50d1							
Preferred	2001:**0**db8:aaaa:0001:**0000**:**0000**:**0000**:**0**200							
Leading 0s omitted	2001:	db8:aaaa:	1:	0:	0:	0: 200		
(::) All-0s segments	2001:	db8:aaaa:	1:: 200					
Compressed	2001:db8:aaaa:1::200							

Prefix Length Notation

In IPv4, the prefix of the address (the network portion) can be represented either by a dotted-decimal netmask (the subnet mask) or through CIDR notation. When we see 192.168.100.0 255.255.255.0 or 192.168.100.0/24, we know that the network portion of the address is the first 24 bits of the address (192.168.100) and that the last 8 bits (.0) are host bits. IPv6 address prefixes are represented in much the same way as IPv4 address prefixes are written in CIDR notation. IPv6 prefixes are represented using the following format:

`IPv6-Address/Prefix-Length`

The *prefix-length* is a decimal value showing the number of leftmost contiguous bits of the address. It identifies the prefix (the network portion) of the address. In unicast addresses, it is used to separate the prefix portion from the Interface ID. The Interface ID is equivalent to the host portion of an IPv4 address.

Looking at the address

```
2001:db8:aaaa:1111::100/64
```

we know that the leftmost 64 bits are the prefix (network portion) and the remaining bits are the Interface ID (host portion). See Figure 5-2.

Each hexadecimal digit is 4 bits; a hextet is a 16-bit segment.

```
2001:db8:aaaa:1111::100/64
```

```
2001 : 0db8 : aaaa : 1111 : 0000 : 0000 : 0000 : 0100
```

| 16 Bits | 16 Bits | 16 Bits | 16 Bits | 16 Bits | 16 Bits | 16 Bits | 16 Bits |

Prefix Length = 64 Bits | Interface ID = 64 Bits

Figure 5-2 IPv6 Prefix and Prefix Length

A /64 prefix length results in an Interface ID of 64 bits. This is a common prefix length for most end-user networks. A /64 prefix length gives us 2^{64} or 18 quintillion devices on a single network (or subnet).

There are several more common prefix length examples, as shown in Figure 5-3. All of these examples fall either on a hextet boundary or on a nibble boundary (a multiple of 4 bits). Although prefix lengths do not need to fall on a nibble boundary, most usually do.

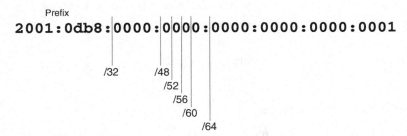

Prefix
2001:0db8:0000:0000:0000:0000:0000:0001

/32   /48
      /52
      /56
      /60
      /64

Figure 5-3 IPv6 Prefix Length Examples

IPv6 Address Types

In IPv6, there are three types of addresses: unicast, multicast, and anycast. This section gives a (very) high-level overview of these types.

NOTE: IPv6 does not have a broadcast address. There are other options that exist in IPv6 that deal with this issue, but this is beyond the scope of this book.

Figure 5-4 diagrams the three types of addresses.

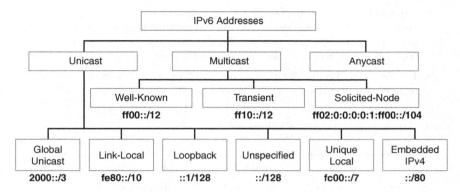

Figure 5-4 IPv6 Address Types

Unicast Addresses

A unicast address uniquely identifies an interface on an IPv6 device. A packet sent to a unicast address is received by the interface that is assigned to that address, Similar to IPv4, a source IPv6 address must be a unicast address.

As shown in Figure 5-4, there are six different types of unicast addresses:

1. **Global unicast:** A routable address in the IPv6 Internet, similar to a public IPv4 address.

2. **Link-local:** Used only to communicate with devices on the same local link.

3. **Loopback:** An address not assigned to any physical interface that can be used for a host to send an IPv6 packet to itself.

4. **Unspecified address:** Used only as a source address and indicates the absence of an IPv6 address.

5. **Unique local:** Similar to a private address in IPv4 (RFC 1918) and not intended to be routable in the IPv6 Internet. However, unlike RFC 1918 addresses, these addresses are not intended to be statefully translated to a global unicast address. Please see Rick Graziani's book *IPv6 Fundamentals* for a more detailed description of stateful translation.

6. **IPv4 embedded:** An IPv6 address that carries an IPv4 address in the low-order 32 bits of an IPv6 address.

Global Unicast Addresses

Global unicast addresses (GUAs) are globally routable and reachable in the IPv6 Internet. The generic structure of a GUA has three fields:

- **Global Routing Prefix:** The prefix or network portion of the address assigned by the provider, such as an ISP, to the customer site.

- **Subnet ID:** A separate field for allocating subnets within the customer site. Unlike IPv4, it is not necessary to borrow bits from the Interface ID (host portion) to create subnets. The number of bits in the Subnet ID falls between where the Global Routing Prefix ends and the Interface ID begins.

- **Interface ID:** Identifies the interface on the subnet, equivalent to the host portion of an IPv4 address. In most cases, the Interface ID is 64 bits in length.

Figure 5-5 shows the structure of a global unicast address.

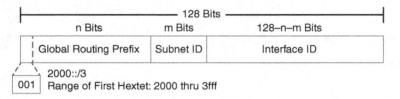

Figure 5-5   Structure of a Global Unicast Address

Link-Local Unicast Addresses

A link-local unicast address is a unicast address that is confined to a single link (a single subnet). Link-local addresses only need to be unique on the link (subnet) and do not need to be unique beyond the link. Therefore, routers do not forward packets with a link-local address.

Figure 5-6 shows the format of a link-local unicast address, which is in the range fe80::/10. Using this prefix and prefix length range results in the range of the first hextet being from fe80 to febf.

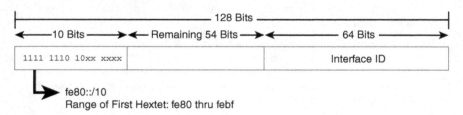

Figure 5-6   Structure of a Link-Local Unicast Address

NOTE: Using a prefix other than fe80 is permitted by RFC 4291, but the addresses should be tested prior to usage.

NOTE: To be an IPv6-enabled device, a device must have an IPv6 link-local address. You do not need to have an IPv6 global unicast address, but you must have a link-local address.

NOTE: Devices dynamically (automatically) create their own link-local IPv6 addresses upon startup. Link-local addresses can be manually configured.

NOTE: Link-local addresses only need to be unique on the link. It is very likely, and even desirable, to have the same link-local address on different interfaces that are on different links. For example, on a device named Router2, you may want all link-local interfaces to be manually configured to FE80::2, whereas all link-local interfaces on Router3 to be manually configured to FE80::3, and so on.

NOTE: There can be only one link-local address per interface. There can be multiple global unicast addresses per interface.

Loopback Addresses

An IPv6 loopback address is ::1, an all-0s address except for the last bit, which is set to 1. It is equivalent to the IPv4 address block 127.0.0.0/8, most commonly the 127.0.0.1 loopback address. The loopback address can be used by a node to send an IPv6 packet to itself, typically when testing the TCP/IP stack.

Table 5-4 shows the different formats for representing an IPv6 loopback address.

TABLE 5-4   IPv6 Loopback Address Representation

Representation	IPv6 Loopback Address
Preferred	0000:0000:0000:0000:0000:0000:0000:0001
Leading 0s omitted	0:0:0:0:0:0:0:1
Compressed	::1

NOTE: A loopback address cannot be assigned to a physical interface.

Unspecified Addresses

An unspecified unicast address is an all-0s address (see Table 5-5), used as a source address to indicate the absence of an address.

Table 5-5 shows the different formats for representing an IPv6 unspecified address.

TABLE 5-5   IPv6 Unspecified Address Representation

Representation	IPv6 Unspecified Address
Preferred	0000:0000:0000:0000:0000:0000:0000:0000
Leading 0s omitted	0:0:0:0:0:0:0:0
Compressed	::

NOTE: An unspecified address cannot be assigned to a physical interface.

Unique Local Addresses

Figure 5-7 shows the structure of the unique local address (ULA), which is the counterpart of IPv4 private addresses. ULAs are used similarly to global unicast addresses, but are for private use and cannot be routed in the global Internet. ULAs are defined in RFC 4193.

Figure 5-7 shows the different formats for representing an IPv6 unspecified address.

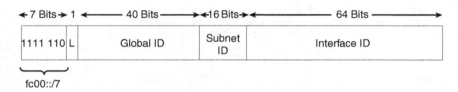

Figure 5-7   Structure of a Unique Local Unicast Address

IPv4 Embedded Addresses

Figure 5-8 shows the structure of IPv4 embedded addresses. They are used to aid in the transition from IPv4 to IPv6. IPv4 embedded addresses carry an IPv4 address in the low-order 32 bits of an IPv6 address.

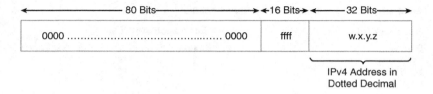

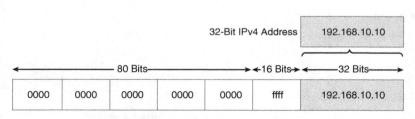

IPv6 Compressed Format   ::ffff:192.168.10.10

Figure 5-8   IPv4-Mapped IPv6 Address

NOTE: This is a transition technique for moving from IPv4 to IPv6 addressing. This should not be used as a permanent solution. The end goal should always be native end-to-end IPv6 connectivity.

Multicast Addresses

Multicast is a technique in which a device sends a single packet to multiple destinations simultaneously (one-to-many transmission). Multiple destinations can actually be multiple interfaces on the same device, but they are typically different devices.

An IPv6 multicast address defines a group of devices known as a multicast group. IPv6 addresses use the prefix ff00::/8, which is equivalent to the IPv4 multicast address 224.0.0.0/4. A packet sent to a multicast group always has a unicast source address; a multicast address can never be the source address.

Unlike IPv4, there is no broadcast address in IPv6. Instead, IPv6 uses multicast.

Table 5-6 shows IPv6 multicast address representation.

TABLE 5-6 IPv6 Multicast Address Representation

Representation	IPv6 Multicast Address
Preferred	ff00:0000:0000:0000:0000:0000:0000:0000/8
Leading 0s omitted	ff00:0:0:0:0:0:0:0/8
Compressed	ff00::/8

The structure of an IPv6 multicast is shown in Figure 5-9; the first 8 bits are 1-bits (ff) followed by 4 bits for flags and a 4-bit Scope field. The next 112 bits represent the Group ID.

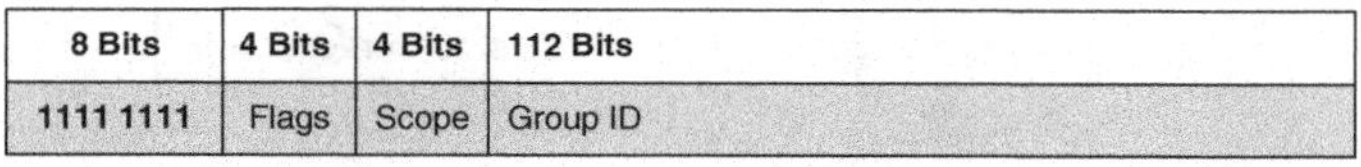

8 Bits	4 Bits	4 Bits	112 Bits
1111 1111	Flags	Scope	Group ID

Figure 5-9 IPv6 Multicast Address

Although there are many different types of multicast addresses, this book defines only two of them:

- Well-known multicast addresses
- Solicited-node multicast addresses

Well-Known Multicast Addresses

Well-known multicast addresses have the prefix ff00::/12. Well-known multicast addresses are predefined or reserved multicast addresses for assigned groups of devices. These addresses are equivalent to IPv4 well-known multicast addresses in the range 224.0.0.0 to 239.255.255.255. Some examples of IPv6 well-known multicast addresses include the following:

Address	Use
ff02::1	All IPv6 devices
ff02::2	All IPv6 routers
ff02::5	All OSPFv3 routers
ff02::6	All OSPFv3 DR routers
ff02::9	All RIPng routers
ff02:a	All EIGRPv6 routers
ff02::1:2	All DHCPv6 servers and relay agents

Solicited-Node Multicast Addresses

Solicited-node multicast addresses are used as a more efficient approach to IPv4's broadcast address. A more detailed description is beyond the scope of this book.

Anycast Addresses

An IPv6 anycast address is an address that can be assigned to more than one interface (typically on different devices). In other words, multiple devices can have the same anycast address. A packet sent to an anycast address is routed to the "nearest" interface having that address, according to the router's routing table.

Figure 5-10 shows an example of anycast addressing.

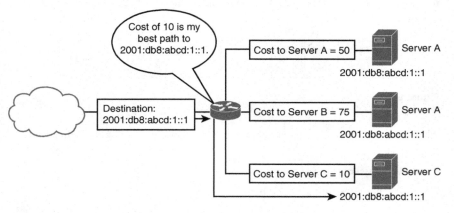

Figure 5-10   Example of Anycast Addressing

NOTE: IPv6 anycast addressing is still somewhat in the experimental stages and beyond the scope of this book.

CHAPTER 6

Cables and Connections

This chapter provides information and commands concerning the following topics:

- Connecting a rollover cable to your router or switch
- Using a USB cable to connect to your router or switch
- Terminal Settings
- LAN Connections
- Serial Cable Types
- Which Cable To Use?
- ANSI/TIA cabling standards
  - T568A versus T568B cables

Connecting a Rollover Cable to Your Router or Switch

Figure 6-1 shows how to connect a rollover cable from your PC to a router or switch.

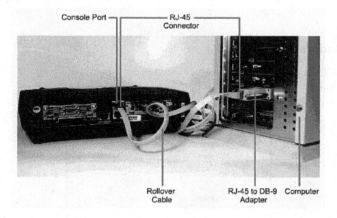

Figure 6-1 Rollover Cable Connection

Using a USB Cable to Connect to Your Router or Switch

On newer Cisco devices, a USB serial console connection is also supported. A USB cable (USB type A to 5-pin mini type B) and operating system driver are needed to establish connectivity. Figure 6-2 shows a Cisco device that can use either a mini-USB connector or a traditional RJ-45 connector.

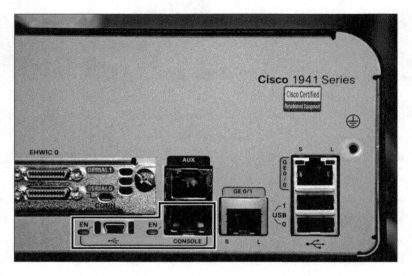

Figure 6-2 Different Console Port Connections

NOTE: Only one console port can be active at a time. If a cable is plugged into the USB port, the RJ-45 port becomes inactive.

NOTE: The OS driver for the USB cable connection is available on the Cisco.com website.

Terminal Settings

Figure 6-3 illustrates the settings that you should configure to have your PC connect to a router or switch.

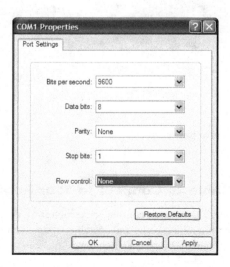

Figure 6-3 PC Settings to Connect to a Router or Switch

LAN Connections

Table 6-1 shows the various port types and connections between LAN devices.

TABLE 6-1 LAN Connections

Port or Connection	Port Type	Connected To	Cable
Ethernet	RJ-45	Ethernet switch	RJ-45
T1/E1 WAN	RJ-48C/CA81A	T1 or E1 network	Rollover
Console	8 pin	Computer COM port	Rollover
Console	USB	Computer USB port	USB
AUX	8 pin	Modem	RJ-45

Serial Cable Types

NOTE: As I mentioned in Chapter 3, "Variable Length Subnet Masking (VLSM)," most networks today no longer use serial connections between routers; Ethernet connections are faster and are now the de facto standards for interconnecting devices. I use serial connections in my *Portable Command Guides* to differentiate between point-to-point connections and LAN connections. Places that use older equipment for their hand-on labs may use serial connections. While not part of the new CCNA 200-301 exam objectives, I am showing this information in this guide to assist those readers using older equipment as part of their learning experience.

Figure 6-4 shows the DB-60 end of a serial cable that connects to a 2500 series router.

Figure 6-4 Serial Cable (2500 Series)

Figure 6-5 shows the newer smart serial end of a serial cable that connects to a smart serial port on your router. Smart serial ports are found on modular routers, such as the newest ISR series (4xxx), older ISR series (x9xx), (x8xx), or on older modular routers such as the 17xx or 26xx series. Smart serial ports on modular routers have to be purchased and installed separately, and are not part of the baseline configuration. Figure 6-2 shows a smart serial card installed into the EHWIC 0 slot on a 1941 series router.

Figure 6-5   Smart Serial Cable (17xx, 26xx, ISR 18xx, 19xx, 28xx, 29xx, 4xxx)

Figure 6-6 shows examples of the male DTE and the female DCE ends that are on the other side of a serial or smart serial cable.

Figure 6-6   V.35 DTE and DCE Cables

Most laptops available today come equipped with USB ports, not serial ports. For these laptops, you need a USB-to-serial connector, as shown in Figure 6-7.

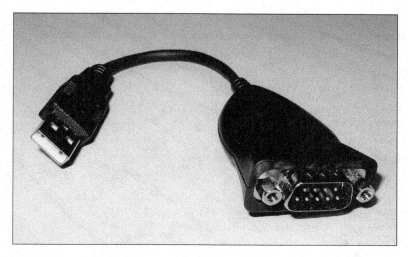

Figure 6-7 USB-to-Serial Connector for Laptops

Figure 6-8 shows an example of a USB type A to 5-pin mini type B cable used for connecting a PC/laptop with a USB port to the USB port on a Cisco device.

Figure 6-8 USB Type A to 5-Pin Mini Type B Cable

Which Cable to Use?

Table 6-2 describes which cable should be used when wiring your devices together. It is important to ensure you have proper cabling; otherwise, you might be giving yourself problems before you even get started.

TABLE 6-2   Determining Which Cables to Use When Wiring Devices Together

If Device A Has a:	And Device B Has a:	Then Use This Cable:
Computer COM/Serial port	RJ-45 Console of router/switch	Rollover
Computer USB port	USB Console of router/switch	USB type A to 5-pin mini type B with appropriate OS drivers
Computer NIC	Switch	Straight-through
Computer NIC	Computer NIC	Crossover
Switch port	Router's Ethernet port	Straight-through
Switch port	Switch port	Crossover (see CAUTION below table)
Router's Ethernet port	Router's Ethernet port	Crossover
Computer NIC	Router's Ethernet port	Crossover
Router's serial port	Router's serial port	Cisco serial DCE/DTE cables

CAUTION: Most switches now have the ability to autosense which type of cable is plugged into a port, and will make the appropriate connection through software to ensure communication occurs. This autosensing of cable types can be disabled. It is therefore recommended to use a crossover cable when connecting two switches via Ethernet ports, to ensure the path of communication is created.

Table 6-3 lists the pinouts for straight-through, crossover, and rollover cables.

TABLE 6-3   Pinouts for Different Cables

Straight-Through Cable	Crossover Cable	Rollover Cable
Pin 1 – Pin 1	Pin 1 – Pin 3	Pin 1 – Pin 8
Pin 2 – Pin 2	Pin 2 – Pin 6	Pin 2 – Pin 7
Pin 3 – Pin 3	Pin 3 – Pin 1	Pin 3 – Pin 6
Pin 4 – Pin 4	Pin 4 – Pin 4	Pin 4 – Pin 5
Pin 5 – Pin 5	Pin 5 – Pin 5	Pin 5 – Pin 4
Pin 6 – Pin 6	Pin 6 – Pin 2	Pin 6 – Pin 3
Pin 7 – Pin 7	Pin 7 – Pin 7	Pin 7 – Pin 2
Pin 8 – Pin 8	Pin 8 – Pin 8	Pin 8 – Pin 1

ANSI/TIA Cabling Standards

ANSI/TIA-568 is a set of telecommunications standards that addresses commercial building cabling for telecommunications products and services. The current standard is revision D (2017), which replaced the 2009 revision C, the 2001 revision B, the 1995 revision A, and the initial issue of T568 in 1991. All of these past revisions and the initial issue are now obsolete.

T568A Versus T568B Cables

One of the most known and discussed features of the ANSI/TIA-568 standard is the definition of the pin assignments, or *pinouts*, on cabling. Pinouts are important because cables do not function properly if the pinouts at the two ends of a cable are not matched correctly. There are two different pinout assignments: T568A and T568B.

The difference between these two standards is pin assignments, not in the use of the different colors (see Table 6-4). The T568A standard is more compatible with voice connections and the Universal Service Order Codes (USOC) standard for telephone infrastructure in the United States. In both T568A and USOC standards, the blue and orange pairs are now on the center four pins; therefore, the colors match more closely with T568A than with the T568B standard.

So, which one is preferred? Information from the standards bodies on this matter is sketchy at best. T568B was traditionally widespread in the United States, whereas places such as Canada and Australia use a lot of T568A. However, T568A is now becoming more dominant in the United States, too.

There is no hard-and-fast rule about which standard to use. It will usually depend on the company you work for and their preferences. Just be consistent. If you are on a job site that uses the B standard, then continue to use B; as I previously mentioned, the A standard more closely aligns with traditional telephone infrastructure, so some installers prefer the A standard.

TABLE 6-4   UTP Wiring Standards

T568A Standard				T568B Standard			
Pin	Color	Pair	Description	Pin	Color	Pair	Description
1	White/green	3	RecvData +	1	White/orange	2	TxData +
2	Green	3	RecvData −	2	Orange	2	TxData −
3	White/orange	2	TxData +	3	White/green	3	RecvData +
4	Blue	1	Unused	4	Blue	1	Unused
5	White/blue	1	Unused	5	White/blue	1	Unused
6	Orange	2	TxData −	6	Green	3	RecvData −
7	White/brown	4	Unused	7	White/brown	4	Unused
8	Brown	4	Unused	8	Brown	4	Unused

TIP: Odd pin numbers are always the striped wires.

TIP: A straight-through (patch) cable is one with both ends using the same standard (T568A or T568B). A crossover cable is one that has T568A on one end and T568B on the other end.

TIP: A rollover (console) cable (used in connecting a PC/laptop to a console port on a Cisco device) is created if you flip the connector (the modular plug) upside down on one side. I have seen many students create a rollover cable in my classes; unfortunately, the assignment was to create a straight-through patch cable, so one end had to be cut off and a new modular plug had to be aligned properly and crimped.

CHAPTER 7

The Command-Line Interface

This chapter provides information and commands concerning the following topics:

- Shortcuts for entering commands
- Using the (Tab⇄) key to enter complete commands
- Console error messages
- Using the question mark for help
- **enable** command
- **exit** command
- **end** command
- **disable** command
- **logout** command
- Setup mode
- Keyboard help
- History commands
- **terminal** commands
- **show** commands
- Using the pipe parameter (|) with the **show** or **more** commands
- Using the **no** and **default** forms of commands

Shortcuts for Entering Commands

To enhance efficiency, Cisco IOS Software has some shortcuts for entering commands. Although these are great to use in the real world, when it comes time to take the CCNA 200-301 exam, make sure you know the full commands, not just the shortcuts.

Router> **enable** = Router> **enab** = Router> **en**	Entering a shortened form of a command is sufficient as long as there is no confusion about which command you are attempting to enter
Router# **configure terminal** = Router# **config t**	

Using the `Tab⇆` Key to Complete Commands

When you are entering a command, you can use the `Tab⇆` key to complete the command. Enter the first few characters of a command and press the `Tab⇆` key. If the characters are unique to the command, the rest of the command is entered in for you. This is helpful if you are unsure about the spelling of a command.

Router# **sh** `Tab⇆` = Router# **show**	

TIP: If your keyboard does not have a `Tab⇆` key, press `Ctrl`-`L` instead.

Console Error Messages

You may see three types of console error messages when working in the CLI:

- Ambiguous command
- Incomplete command
- Invalid input

Error Message	Meaning	What to Do
% Ambiguous Command: "show con"	Not enough characters were entered to allow device to recognize the command.	Reenter the command with a question mark (?) immediately after the last character: **show con?** All possible keywords will be displayed
% Incomplete Command	More parameters need to be entered to complete the command.	Reenter the command followed by a question mark (?). Include a space between the command and the question mark (?)
% Invalid input detected at ^ marker	The command entered has an error. The ^ marks the location of the error.	Reenter the command, correcting the error at the location of the ^. If you are unsure what the error is, reenter the command with a question mark (?) at the point of the error to display the commands or parameters available

Using the Question Mark for Help

The following output shows you how using the question mark can help you work through a command and all its parameters.

Router# **?**	Lists all commands available in the current command mode
Router# **c?** calendar call-home cd clear clock cns configure connect copy crypto	Lists all the possible choices that start with the letter *c*

`Router# cl?`
 	
`clear clock`
	Lists all the possible choices that start with the letters *cl*

`Router# clock`
 	
`% Incomplete Command`
	Tells you that more parameters need to be entered

`Router# clock ?`
 	
`read-calendar`
 	
`set`
 	
`update-calendar`
	Shows all subcommands for this command

`Router# clock set 19:50:00 14` `July 2019 ?` ⏎Enter	Pressing the ⏎Enter key confirms the time and date configured
`Router#`	No Error message/Incomplete command message means the command was entered successfully

enable Command

`Router> enable` `Router#`	Moves the user from user mode to privileged EXEC mode. Notice the prompt changes from > to #

exit Command

`Router# exit`
or
`Router> exit`
	Logs a user off

`Router(config-if)# exit` `Router(config)#`	Moves you back one level
`Router(config)# exit` `Router#`	Moves you back one level

end Command

`Router(config-if)# end` `Router#`	Moves you from the current mode all the way down to privileged EXEC mode. This example moves from interface configuration mode down to privileged EXEC mode. Notice the prompt changes from Router(config-if)# to Router#

disable Command

`Router# disable` `Router>`	Moves you from privileged EXEC mode back to user mode

logout Command

Router# **logout**	Performs the same function as **exit**

Setup Mode

Setup mode starts automatically if there is no startup configuration present.

Router# **setup**	Enters the System Configuration Dialog from the command line

NOTE: The answer inside the square brackets, [ ], is the default answer. If this is the answer you want, just press ⏎Enter. Pressing Ctrl-C at any time will end the setup process, shut down all interfaces, and take you to user mode (Router>).

NOTE: You *cannot* use the System Configuration Dialog (setup mode) to configure an entire router. It does only the basics. For example, you can only turn on RIPv1, but not Open Shortest Path First Protocol (OSPF) or Enhanced Interior Gateway Routing Protocol (EIGRP). You cannot create access control lists (ACLs) here or enable Network Address Translation (NAT). You can assign an IP address to an interface but not to a subinterface. All in all, setup mode is very limiting.

Entering the System Configuration Dialog is not a recommended practice. Instead, you should use the command-line interface (CLI), which is more powerful:

   Would you like to enter the initial configuration dialog? [yes]: **no**

   Would you like to enable autoinstall? [yes]: **no**

Autoinstall is a feature that tries to broadcast out all interfaces when attempting to find a configuration. If you answer **yes**, you must wait for a few minutes while it looks for a configuration to load. Very frustrating. Answer **no**.

Keyboard Help

The keystrokes in the following table are meant to help you edit the configuration. Because you'll want to perform certain tasks again and again, Cisco IOS Software provides certain keystroke combinations to help make the process more efficient.

^ Router# **confog t** ^ % Invalid input detected at '^' marker. Router# **config t** Router(config)#	Shows you where you made a mistake in entering a command
$	Indicates that the line has been scrolled to left
Ctrl-A	Moves cursor to beginning of line
Esc-B	Moves cursor back one word
Ctrl-B (or ←)	Moves cursor back one character

Ctrl-E	Moves cursor to end of line
Ctrl-F (or →)	Moves cursor forward one character
Esc-F	Moves cursor forward one word
Ctrl-◆Shift-6	Allows the user to interrupt an IOS process such as ping or traceroute
Ctrl-Z	Moves you from any prompt back down to privileged EXEC mode. Can also be used to interrupt the output being displayed and return to privileged EXEC mode
Ctrl-W	Deletes the word to the left of the cursor
Ctrl-U	Deletes the entire line
Ctrl-T	Swaps or transposes the current character with the one before it
Ctrl-K	Erases characters from the cursor to the end of the line
Ctrl-X	Erases characters from the cursor to the beginning of the line
Ctrl-D	Deletes from cursor to the end of the word
Delete	Removes characters to the right of the cursor
←Backspace	Removes characters to the left of the cursor
Ctrl-L	Reprints the line
Ctrl-R	Refreshes the line—use this if the system sends a message to the screen while a command is being entered and you are not using line synchronization. This brings your command to the next line without the message interfering with the command
Ctrl-C	Exits from global configuration mode and moves you to privileged EXEC mode
Esc-C	Makes the letter at the cursor uppercase
Esc-L	Makes the letter at the cursor lowercase
Esc-U	Makes the letters from the cursor to the end of the word uppercase
Router# **terminal no editing** Router#	Turns off the ability to use the previous keyboard shortcuts
Router# **terminal editing** Router#	Reenables enhanced editing mode (can use previous keyboard shortcuts)

History Commands

Ctrl-P (or ↑)	Recalls commands in the history buffer in a backward sequence, beginning with the most recent command
Ctrl-N (or ↓)	Returns to more recent commands in the history buffer after recalling commands with the Ctrl-P key sequence

terminal Commands

Router# **terminal** **no editing** Router#	Turns off the ability to use keyboard shortcuts
Router# **terminal** **editing** Router#	Reenables enhanced editing mode (can use keyboard shortcuts)
Router# **terminal** **length** x	Sets the number of lines displayed in a **show** command to x, where x is a number between 0 and 512 (the default is 24)

NOTE: If you set the **terminal length** x command to zero (0), the router will not pause between screens of output.

Router# **terminal history** **size** *number*	Sets the number of commands in the buffer that can be recalled by the router (maximum 256)
Router# **terminal history** **size 25**	Causes the router to remember the last 25 commands in the buffer
Router# **no terminal history** **size 25**	Sets the history buffer back to 10 commands, which is the default

NOTE: The **history size** command provides the same function as the **terminal history size** command.

Be careful when you set the size to something larger than the default. By telling the router to keep the last 256 commands in a buffer, you are taking memory away from other parts of the router. What would you rather have: a router that remembers what you last typed in or a router that routes as efficiently as possible?

show Commands

Router# **show version**	Displays information about the current Cisco IOS Software
Router# **show flash**	Displays information about flash memory
Router# **show history**	Lists all commands in the history buffer

NOTE: The last line of output from the **show version** command tells you what the configuration register is set to.

Using the Pipe Parameter (|) with the show or more Commands

By using a pipe (|) character in conjunction with a **show** command or a **more** command, you can set filters for specific information that you are interested in.

Router# **show running-config | include hostname**	Displays configuration information that includes the specific word *hostname*
Router# **show running-config | section FastEthernet 0/1**	Displays configuration information about the section FastEthernet 0/1

The Pipe Parameter (l) Options Parameter	The Pipe Parameter (l) Options Description
begin	Shows all output from a certain point, starting with the line that matches the filtering expression
Router# **show running-config | begin line con 0**	Output begins with the first line that has the regular expression "line con 0"
exclude	Excludes all output lines that match the filtering expression
Router# **show running-config | exclude interface**	Any line with the regular expression "interface" will not be shown as part of the output
include	Includes all output lines that match the filtering expression
Router# **show running-config | include duplex**	Any line that has the regular expression "duplex" will be shown as part of the output
Router# **show running-config | include (is)**	Displays only lines that contain the regular expression (is). The parentheses force the inclusion of the spaces before and after "is". This ensures that lines containing "is" with a space before and after it will be included in the output, and output without spaces will be excluded—words like "disconnect" or "isdn" will be excluded
section	Shows the entire section that starts with the filtering expression
Router# **show running-config | section interface GigabitEthernet 0/0**	Displays information about interface GigabitEthernet 0/0
Router# **more nvram: startup-config | begin ip**	Displays output from the startup-config file that begins with the first line that contains the regular expression "ip"
Router# **more nvram: startup-config | include ip**	Displays output from the startup-config file that only includes the regular expression "ip"

NOTE: You can use the pipe parameter and filters with any **show** command.

NOTE: The filtering expression has to match **exactly** with the output you want to filter. You cannot use shortened forms of the items you are trying to filter. For example, the command

```
Router# show running-config | section gig0/0
```

will not work because there is no section in the running-config called gig0/0. You must use the expression GigabitEthernet0/0 with no spelling errors or extra spaces added in.

Using the no and default Forms of Commands

Almost every configuration command has a **no** form. In general, use the **no** form to disable a feature or function. Use the command without the **no** keyword to reenable a disabled feature or to enable a feature that is disabled by default.

`Router(config)#` **`router`** **`eigrp 100`**	Enables the EIGRP routing process with Autonomous System number 100
`Router(config)#` **`no router`** **`eigrp 100`**	Disables EIGRP routing process 100 and removes the entire EIGRP configuration from the running configuration
`Router(config)#` **`no ip routing`**	Disables IP routing on the device (IP routing is enabled by default)
`Router(config)#` **`ip routing`**	Reenables IP routing on the device

Many CLI commands also have a default form. By issuing the **default** *command-name* command, you can configure the command to its default setting. The Cisco IOS Software command reference documents located on Cisco.com describe the function of the **default** form of the command when it performs a different function from either the plain form or the **no** form of the command. To see what default commands are available on your system, enter **default ?** in the appropriate command mode:

`Router(config)#` **`default ?`**	Lists all commands that are available at this mode for use with the **default** command

CHAPTER 8

Configuring a Switch

This chapter provides information and commands concerning the following topics:

- Help commands
- Command modes
- Verifying commands
- Resetting switch configuration
- Setting host names
- Setting passwords
- Setting IP addresses and default gateways
- Setting interface descriptions
- The **mdix auto** command
- Setting duplex operation
- Setting operation speed
- Managing the MAC address table
- Configuration example

NOTE: As of the publication of this Command Guide, the latest switch platform from Cisco is the Catalyst 9xxx series. This series consists of wireless APs, Layer 2 switches, and Layer 3 switches for use in small branch deployments up to large core deployments. For those using older Catalyst switches, this book makes the following assumptions:

- Layer 2 switches are named either *Switch2960* or *Switch9200*.
- Layer 3 switches are named either *Switch3650* or *Switch9300*.
- If you have any questions regarding the validity of commands on your devices, refer to your device-specific/OS-specific documentation on the Cisco.com website.

Help Commands

`switch> ?`	Lists all commands available in the current command mode **TIP:** The **?** works here the same as in a router

`switch# c?`
`cd clear clock cns`
`configure`
`connect copy`
	Lists all the possible choices that start with the letter *c*

`switch# show ?`	Shows all parameters for this command

Command Modes

`switch> enable`	Moves the user from user mode to privileged mode **TIP:** This is the same command as used in a router
`switch#`	Indicates privileged mode
`switch# disable`	Leaves privileged mode
`switch> exit`	Leaves user mode

Verifying Commands

`switch# show version`	Displays information about software and hardware
`switch# show flash:`	Displays information about flash memory
`switch# show mac` `address-table`	Displays the current MAC address forwarding table
`switch# show controllers` `ethernet-controller`	Displays information about the Ethernet controller
`switch# show running-config`	Displays the current configuration in DRAM
`switch# show startup-config`	Displays the current configuration in NVRAM
`switch# show post`	Displays whether the switch passed POST
`switch# show vlan`	Displays the current VLAN configuration
`switch# show interfaces`	Displays the interface configuration and status of line: up/up, up/down, admin down **NOTE:** The **show interfaces** command is unsupported in some earlier Cisco IOS Software releases, such as 12.2(25)FX
`switch# show interfaces` `vlan1`	Displays setting of virtual interface VLAN 1, the default VLAN on the switch **NOTE:** The **show interfaces vlanx** command is unsupported in some earlier Cisco IOS Software releases, such as 12.2(25)FX

Resetting Switch Configuration

Switch# **delete flash:vlan.dat**	Removes the VLAN database from flash memory
Delete filename [vlan.dat]?	Press ↵Enter
Delete flash:vlan.dat? [confirm]	Reconfirm by pressing ↵Enter
Switch# **erase startup-config**	Erases the file from NVRAM
\<output omitted>	
Switch# **reload**	Restarts the switch
Switch# **write erase**	Erases the startup-config file from NVRAM. This is an older version of the **erase startup-config** command

Setting Host Names

Switch# **configure terminal**	Moves to global configuration mode
Switch(config)# **hostname Switch9200**	Creates a locally significant host name of the switch. This is the same command as is used on routers
Switch9200(config)#	

TIP: If you set a host name that begins with a number, you receive a warning about using illegal characters. However, the switch accepts the name.

```
Switch(config)# hostname 9200
% Hostname contains one or more illegal characters.
9200(config)#
```

Setting Passwords

Setting passwords for the 2960/9200 series switches is the same method as used for a router.

Switch9200(config)# **enable password cisco**	Sets the enable password to *cisco*
Switch9200(config)# **enable secret class**	Sets the encrypted secret password to *class*
Switch9200(config)# **line console 0**	Enters line console mode
Switch9200(config-line)# **login**	Enables password checking
Switch9200(config-line)# **password cisco**	Sets the password to *cisco*
Switch9200(config-line)# **exit**	Exits line console mode
Switch9200(config-line)# **line vty 0 15**	Enters line vty mode for all 15 virtual ports

Switch9200(config-line)# **login**	Enables password checking
Switch9200(config-line)# **password cisco**	Sets the password to *cisco*
Switch9200(config-line)# **exit**	Exits line vty mode
Switch9200(config)#	

Setting IP Addresses and Default Gateways

Switch9200(config)# **interface vlan1**	Enters the virtual interface for VLAN 1, the default VLAN on the switch
Switch9200(config-if)# **ip address 172.16.10.2 255.255.255.0**	Sets the IP address and netmask to allow for remote access to the switch
Switch9200(config-if)# **exit**	Returns to global configuration mode
Switch9200(config)# **ip default-gateway 172.16.10.1**	Allows IP information an exit past the local network

> **TIP:** For the 2960/9200 series switches, the IP address of the switch is just that—the IP address for the *entire* switch. That is why you set the address in VLAN 1 (the default VLAN of the switch) and not in a specific Ethernet interface. If you choose to make your management VLAN a different number, you would use these commands in that VLAN using the *interface vlan x* command, where *x* is the number of your management VLAN.

Setting Interface Descriptions

Switch2960(config)# **interface fastethernet 0/1**	Enters interface configuration mode
Switch2960(config-if)# **description Finance VLAN**	Adds a description of the interface. The description is locally significant only

> **TIP:** The 2960 series switches have ports ranging from 8 to 48 Fast Ethernet ports named fa0/1, fa0/2, ... fa0/48—there is no fastethernet 0/0. This is true for the 2960G series, in which all ports are Gigabit Ethernet ports named gi0/1, gi0/2 ... gi0/48. Again, there is no gigabitethernet0/0 port.

The 9200 series switches have GigabitEthernet (GE) and 10-GE ports only.

The mdix auto Command

When automatic medium-dependent interface crossover (Auto-MDIX) is enabled on an interface, the interface automatically detects the required cable connection type (straight-through or crossover) and configures the connection appropriately. When connecting switches without the Auto-MDIX feature, you must use straight-through cables to connect to devices such as servers, workstations, or routers and use crossover cables to connect to other switches or repeaters. With Auto-MDIX enabled, you can use either type of cable to connect to other devices, and the interface automatically corrects for incorrect cabling.

`Switch2960(config)# `**`interface`** `fastethernet 0/1`	Enters interface configuration mode
`Switch2960(config-if)# `**`mdix auto`**	Enables Auto-MDIX on the interface
`Switch2960(config-if)# `**`no mdix auto`**	Disables Auto-MDIX on the interface

TIP: The Auto-MDIX feature is enabled by default on switches running Cisco IOS Release 12.2(18)SE or later. For releases between Cisco IOS Release 12.1(14)EA1 and 12.2(18)SE, the Auto-MDIX feature is disabled by default.

TIP: If you are working on a device where Auto-MDIX is enabled by default, the command does *not* show up when you enter **show running-config**.

CAUTION: When you enable Auto-MDIX, you must also set the interface speed and duplex to **auto** so that the feature operates correctly. In other words, if you use Auto-MDIX to give you the flexibility to use either type of cable to connect your switches, you lose the ability to hard-set the speed/duplex on both sides of the link.

The following table shows the different link state results from Auto-MDIX settings with correct and incorrect cabling.

Local Side Auto-MDIX	Remote Side Auto-MDIX	With Correct Cabling	With Incorrect Cabling
On	On	Link up	Link up
On	Off	Link up	Link up
Off	On	Link up	Link up
Off	Off	Link up	Link down

Setting Duplex Operation

`Switch2960(config)# `**`interface`** `fastethernet 0/1`	Moves to interface configuration mode
`Switch2960(config-if)# `**`duplex full`**	Forces full-duplex operation
`Switch2960(config-if)# `**`duplex auto`**	Enables auto-duplex config
`Switch2960(config-if)# `**`duplex half`**	Forces half-duplex operation

Setting Operation Speed

`Switch2960(config)# `**`interface`** `fastethernet 0/1`	Moves to interface configuration mode
`Switch2960(config-if)# `**`speed 10`**	Specifies that the port runs at 10 Mbps
`Switch2960(config-if)# `**`speed 100`**	Specifies that the port runs at 100 Mbps
`Switch9200(config-if)# `**`speed 1000`**	Specifies that the port runs at 1000 Mbps
`Switch9200(config-if)# `**`speed 2500`**	Specifies that the port runs at 2500 Mbps. This option is only valid and visible on multi-Gigabit-supported Ethernet ports

`Switch9200(config-if)# speed 5000`	Specifies that the port runs at 5000 Mbps. This option is only valid and visible on multi-Gigabit-supported Ethernet ports
`Switch2960(config-if)# speed auto`	Detects the speed at which the port should run, automatically, based on the port at the other end of the link
`Switch9200(config-if)# speed nonegotiate`	Disables autonegotiation, and the port runs at 1000 Mbps

Managing the MAC Address Table

`switch# show mac address-table`	Displays current MAC address forwarding table
`switch# clear mac address-table`	Deletes all entries from current MAC address forwarding table
`switch# clear mac address-table dynamic`	Deletes only dynamic entries from table

Configuration Example

Figure 8-1 shows the network topology for the basic configuration of a 2960 series switch using commands covered in this chapter. These commands will also work on a 9200 series switch, with the exception that all Fast Ethernet ports will be Gigabit Ethernet ports on the 9200 switch.

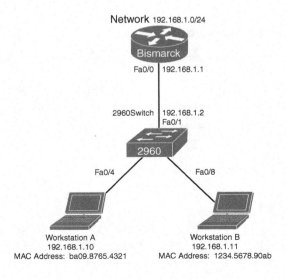

Figure 8-1   Network Topology for 2960 Series Switch Configuration

`switch>` **`enable`**	Enters privileged EXEC mode
`switch#` **`configure terminal`**	Enters global configuration mode
`switch(config)#` **`no ip domain-lookup`**	Turns off Domain Name System (DNS) queries so that spelling mistakes do not slow you down
`switch(config)#` **`hostname Switch2960`**	Sets the host name
`Switch2960(config)#` **`enable secret cisco`**	Sets the encrypted secret password to *cisco*
`Switch2960(config)#` **`line console 0`**	Enters line console mode
`Switch2960(config-line)#` **`logging synchronous`**	Appends commands to a new line; switch information will not interrupt
`Switch2960(config-line)#` **`login`**	User must log in to console before use
`Switch2960(config-line)#` **`password switch`**	Sets the console password to *switch*
`Switch2960(config-line)#` **`exec-timeout 0 0`**	The console line will not log out because of the connection to the console being idle
`Switch2960(config-line)#` **`exit`**	Moves back to global configuration mode
`Switch2960(config)#` **`line vty 0 15`**	Moves to configure all 16 vty ports at the same time
`Switch2960(config-line)#` **`login`**	User must log in to vty port before use
`Switch2960(config-line)#` **`password class`**	Sets the vty password to *class*
`Switch2960(config-line)#` **`exit`**	Moves back to global configuration mode
`Switch2960(config)#` **`ip default-gateway 192.168.1.1`**	Sets default gateway address
`Switch2960(config)#` **`interface vlan 1`**	Moves to virtual interface VLAN 1 configuration mode
`Switch2960(config-if)#` **`ip address 192.168.1.2 255.255.255.0`**	Sets the IP address and netmask for switch
`Switch2960(config-if)#` **`no shutdown`**	Turns the virtual interface on
`Switch2960(config-if)#` **`interface fastethernet 0/1`**	Moves to interface configuration mode for fastethernet 0/1
`Switch2960(config-if)#` **`description Link to Bismarck Router`**	Sets a local description
`Switch2960(config-if)#` **`interface fastethernet 0/4`**	Moves to interface configuration mode for fastethernet 0/4
`Switch2960(config-if)#` **`description Link to Workstation A`**	Sets a local description
`Switch2960(config-if)#` **`interface fastethernet 0/8`**	Moves to interface configuration mode for fastethernet 0/8

`Switch2960(config-if)#` **`description`** **`Link to Workstation B`**	Sets a local description
`Switch2960(config-if)#` **`exit`**	Returns to global configuration mode
`Switch2960(config)#` **`exit`**	Returns to privileged EXEC mode
`Switch2960#` **`copy running-config`** **`startup-config`**	Saves the configuration to NVRAM
`Switch2960#`	

This chapter provides information and commands concerning the following topics:

- Creating static VLANs
    - Creating static VLANs using VLAN configuration mode
- Assigning ports to VLANs
- Using the **range** command
- Configuring a voice VLAN
    - Configuring voice and data with trust
    - Configuring voice and data without trust
- Verifying VLAN information
- Saving VLAN configurations
- Erasing VLAN configurations
- Configuration example: VLANs

Creating Static VLANs

Static VLANs occur when a switch port is manually assigned by the network administrator to belong to a VLAN. Each port is associated with a specific VLAN. By default, all ports are originally assigned to VLAN 1. You create VLANs using the VLAN configuration mode.

Creating Static VLANs Using VLAN Configuration Mode

Switch(config)# **vlan 3**	Creates VLAN 3 and enters VLAN configuration mode for further definitions
Switch(config-vlan)# **name Engineering**	Assigns a name to the VLAN. The length of the name can be from 1 to 32 characters
Switch(config-vlan)# **exit**	Applies changes, increases the revision number by 1, and returns to global configuration mode
Switch(config)#	

NOTE: Use this method to add normal-range VLANs (1–1005) or extended-range VLANs (1006–4094). Configuration information for normal-range VLANs is always saved in the VLAN database, and you can display this information by entering the **show vlan** privileged EXEC command.

NOTE: The VLAN Trunking Protocol (VTP) revision number is increased by one each time a VLAN is created or changed.

VTP version 3 supports propagation of extended-range VLANs. VTP versions 1 and 2 propagate only VLANs 1–1005.

NOTE: Transparent mode does not increment the VTP revision number.

Assigning Ports to VLANs

`Switch(config)# interface fastethernet 0/1`	Moves to interface configuration mode
`Switch(config-if)# switchport mode access`	Sets the port to access mode
`Switch(config-if)# switchport access vlan 10`	Assigns this port to VLAN 10

NOTE: When you use the **switchport mode access** command, the port operates as a nontrunking, single VLAN interface.

TIP: An access port can belong to only one data VLAN.

TIP: By default, all ports are members of VLAN 1.

Using the range Command

`Switch(config)# interface range fastethernet 0/1 - 9`	Enables you to set the same configuration parameters on multiple ports at the same time **NOTE:** Depending on the model of switch, there is a space before and after the hyphen in the **interface range** command. Be careful with your typing
`Switch(config-if-range)# switchport mode access`	Sets ports 1 to 9 as access ports
`Switch(config-if-range)# switchport access vlan 10`	Assigns ports 1 to 9 to VLAN 10

Configuring a Voice VLAN

The voice VLAN feature permits switch ports to carry voice traffic with Layer 3 precedence and Layer 2 Class of Service (CoS) values from an IP Phone.

You can configure the switch port, which is connected to an IP Phone, to use one VLAN for voice traffic and another VLAN for data traffic originating from a device that is connected to the access port of the IP Phone.

Cisco switches use Cisco Discovery Protocol (CDP) packets to communicate with the IP Phone. CDP must be enabled on any switch port that is to be connected to an IP Phone.

NOTE: Voice VLANs are disabled by default.

NOTE: By default, a switch port drops any tagged frames in hardware.

Configuring Voice and Data with Trust

NOTE: This configuration is used for Cisco IP Phones that trust data traffic using CoS coming from the laptop or PC connected to the IP Phone's access port. Data traffic uses the native VLAN.

`Switch# configure terminal`	Enters global configuration mode
`Switch(config)# mls qos`	Enables QoS functionality globally
`Switch(config)# interface fastethernet 0/6`	Moves to interface configuration mode
`Switch(config-if)# mls qos trust cos`	Has the interface enter into a state of trust and classifies traffic by examining the incoming Class of Service (CoS)
`Switch(config-if)# mls qos trust dscp`	Has the interface enter into a state of trust and classifies traffic by examining the incoming Differentiated Services Code Point (DSCP) value
`Switch(config-if)# switchport voice vlan dot1p`	Configures the telephone to use the IEEE 802.1p priority tagging to forward all voice traffic with a higher priority through VLAN 0 (the native VLAN). By default the Cisco IP Phone forwards the voice traffic with an IEEE 802.1p priority of 5
`Switch(config-if)# switchport voice vlan none`	Does not instruct the IP telephone about the voice VLAN. The telephone uses the configuration from the telephone keypad
`Switch(config-if)# switchport voice vlan untagged`	Configures the telephone to send untagged voice traffic. This is the default for the telephone
`Switch(config-if)# switchport voice vlan 10`	Configures voice VLAN 10
`Switch(config-if)# switchport voice vlan 10 name vlan_name`	Optional command. Specifies the VLAN name to be used for voice traffic. You can enter up to 128 characters
`Switch(config-if)# switchport priority extend trust`	Extends the trust state to the device (PC) connected to the access port of the IP Phone The switch instructs the phone on how to process data packets from the device (PC) connected to the IP Phone
`Switch(config-if)# priority-queue out`	Gives voice packets head-of-line privileges when trying to exit the port. This helps prevent jitter
`Switch(config-if)# spanning-tree portfast`	Enables PortFast on the interface, which removes the interface from the Spanning Tree Protocol (STP)
`Switch(config-if)# spanning-tree bpduguard enable`	Enables Bridge Protocol Data Unit (BPDU) Guard on the interface
`Switch(config-if)# exit`	Exits interface configuration mode and returns to global configuration mode
`Switch(config)#`	

Configuring Voice and Data Without Trust

NOTE: This configuration is used for Cisco IP Phones without trusting the laptop or
PC connected to the IP Phone's access port. Data traffic uses the 802.1Q frame type.

Switch# **configure terminal**	Enters global configuration mode
Switch(config)# **mls qos**	Enables QoS functionality globally
Switch(config)# **interface fastethernet 0/8**	Moves to interface configuration mode
Switch(config-if)# **mls qos trust cos**	Has the interface enter into a state of trust and classifies traffic by examining the incoming Class of Service (CoS) value
Switch(config-if)# **mls qos trust dscp**	Has the interface enter into a state of trust and classifies traffic by examining the incoming Differentiated Services Code Point (DSCP) value
Switch(config-if)# **switchport voice vlan 10**	Configures voice VLAN 10
Switch(config-if)# **switchport access vlan 20**	Configures data VLAN 20
Switch(config-if)# **priority-queue out**	Gives voice packets head-of-line privileges when trying to exit the port. This helps prevent jitter
Switch(config-if)# **spanning-tree portfast**	Enables PortFast on the interface, which removes the interface from the Spanning Tree Protocol (STP)
Switch(config-if)# **spanning-tree bpduguard enable**	Enables BPDU Guard on the interface
Switch(config-if)# **exit**	Exits interface configuration mode and returns to global configuration mode
Switch(config)#	

Verifying VLAN Information

Switch# **show vlan**	Displays VLAN information
Switch# **show vlan brief**	Displays VLAN information in brief
Switch# **show vlan id 2**	Displays information about VLAN 2 only
Switch# **show vlan name marketing**	Displays information about VLAN named marketing only
Switch# **show interfaces vlan x**	Displays interface characteristics for the specified VLAN
Switch# **show interfaces switchport**	Displays VLAN information for all interfaces
Switch# **show interfaces fastethernet 0/6 switchport**	Displays VLAN information (including voice VLAN information) for the specified interface

Saving VLAN Configurations

The configurations of VLANs 1 to 1005 are always saved in the VLAN database. When using VLAN configuration mode, the **exit** command saves the changes to the VLAN database.

If the VLAN database configuration is used at startup, and the startup configuration file contains extended-range VLAN configuration, this information is lost when the system boots.

If you are using VTP transparent mode, the configurations are also saved in the running configuration and can be saved to the startup configuration using the **copy running-config startup-config** command.

If the VTP mode is transparent in the startup configuration, and the VLAN database and the VTP domain name from the VLAN database match those in the startup configuration file, the VLAN database is ignored (cleared), and the VTP and VLAN configurations in the startup configuration file are used. The VLAN database revision number remains unchanged in the VLAN database.

Erasing VLAN Configurations

`Switch# ` **`delete flash:`** **`vlan.dat`**	Removes the entire VLAN database from flash
	
	CAUTION: Make sure there is no space between the colon (:) and the characters vlan.dat. You can potentially erase the entire contents of the flash with this command if the syntax is not correct. Make sure you read the output from the switch. If you need to cancel, press Ctrl-C to escape back to privileged mode:
	
	`(Switch#)`
	`Switch# `**`delete flash:vlan.dat`**
	`Delete filename [vlan.dat]?`
	`Delete flash:vlan.dat? [confirm]`
	`Switch#`

`Switch(config)# ` **`interface fastethernet 0/5`**	Moves to interface configuration mode
`Switch(config-if)# ` **`no switchport access vlan 5`**	Removes port from VLAN 5 and reassigns it to VLAN 1—the default VLAN
`Switch(config-if)# ` **`exit`**	Moves to global configuration mode
`Switch(config)# ` **`no vlan 5`**	Removes VLAN 5 from the VLAN database

NOTE: When you delete a VLAN from a switch that is in VTP server mode, the VLAN is removed from the VLAN database for all switches in the VTP domain. When you delete a VLAN from a switch that is in VTP transparent mode, the VLAN is deleted only on that specific switch.

NOTE: You cannot delete the default VLANs for the different media types: Ethernet VLAN 1 and FDDI or Token Ring VLANs 1002 to 1005.

CAUTION: When you delete a VLAN, any ports assigned to that VLAN become inactive. They remain associated with the VLAN (and thus inactive) until you assign them to a new VLAN. Therefore, it is recommended that you reassign ports to a new VLAN or the default VLAN before you delete a VLAN from the VLAN database.

Configuration Example: VLANs

Figure 9-1 illustrates the network topology for the configuration that follows, which shows how to configure VLANs using the commands covered in this chapter.

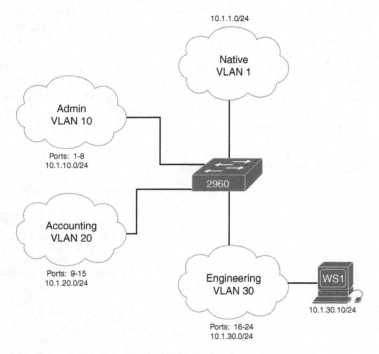

Figure 9-1   Network Topology for VLAN Configuration Example

2960 Switch

Switch> **enable**	Moves to privileged mode
Switch# **configure terminal**	Moves to global configuration mode
Switch(config)# **hostname Switch2960**	Sets the host name
Switch2960(config)# **vlan 10**	Creates VLAN 10 and enters VLAN configuration mode
Switch2960(config-vlan)# **name Admin**	Assigns a name to the VLAN

`Switch2960(config-vlan)# ` **`exit`**	Increases the revision number by 1 and returns to global configuration mode
`Switch2960(config)# ` **`vlan 20`**	Creates VLAN 20 and enters VLAN configuration mode
`Switch2960(config-vlan)# ` **`name`** **`Accounting`**	Assigns a name to the VLAN
`Switch2960(config-vlan)# ` **`vlan 30`**	Creates VLAN 30 and enters VLAN configuration mode. Note that you do not have to exit back to global configuration mode to execute this command. This also increases the revision number by 1 because you moved from VLAN 20 to VLAN 30
`Switch2960(config-vlan)# ` **`name`** **`Engineering`**	Assigns a name to the VLAN
`Switch2960(config-vlan)# ` **`exit`**	Increases the revision number by 1 and returns to global configuration mode
`Switch2960(config)# ` **`interface range`** **`fasthethernet 0/1 - 8`**	Enables you to set the same configuration parameters on multiple ports at the same time
`Switch2960(config-if-range)# ` **`switchport mode access`**	Sets ports 1 to 8 as access ports
`Switch2960(config-if-range)# ` **`switchport access vlan 10`**	Assigns ports 1 to 8 to VLAN 10
`Switch2960(config-if-range)# ` **`interface range`** **`fastethernet 0/9 - 15`**	Enables you to set the same configuration parameters on multiple ports at the same time
`Switch2960(config-if-range)# ` **`switchport mode access`**	Sets ports 9 to 15 as access ports
`Switch2960(config-if-range)# ` **`switchport access vlan 20`**	Assigns ports 9 to 15 to VLAN 20
`Switch2960(config-if-range)# ` **`interface range fastethernet`** **`0/16 - 24`**	Enables you to set the same configuration parameters on multiple ports at the same time
`Switch2960(config-if-range)# ` **`switchport mode access`**	Sets ports 16 to 24 as access ports
`Switch2960(config-if-range)# ` **`switchport access vlan 30`**	Assigns ports 16 to 24 to VLAN 30
`Switch2960(config-if-range)# ` **`exit`**	Returns to global configuration mode
`Switch2960(config)# ` **`exit`**	Returns to privileged mode
`Switch2960# ` **`copy running-config`** **`startup-config`**	Saves the configuration in NVRAM

CHAPTER 10

VLAN Trunking Protocol and Inter-VLAN Communication

This chapter provides information and commands concerning the following topics:

- Dynamic Trunking Protocol (DTP)
- Setting the VLAN encapsulation type
- VLAN Trunking Protocol (VTP)
- Verifying VTP
- Inter-VLAN communication using an external router: router-on-a-stick
- Inter-VLAN communication on a multilayer switch through a switch virtual interface (SVI)
- Removing L2 switchport capability of an interface on an L3 switch
- Configuring Inter-VLAN communication on an L3 switch
- Inter-VLAN communication tips
- Configuration example: Inter-VLAN communication

Dynamic Trunking Protocol (DTP)

`Switch(config)# interface fastethernet 0/1`	Moves to interface configuration mode
`Switch(config-if)# switchport mode dynamic desirable`	Makes the interface actively attempt to convert the link to a trunk link **NOTE:** With the **switchport mode dynamic desirable** command set, the interface becomes a trunk link if the neighboring interface is set to **trunk**, **desirable**, or **auto**
`Switch(config-if)# switchport mode dynamic auto`	Makes the interface able to convert into a trunk link **NOTE:** With the **switchport mode dynamic auto** command set, the interface becomes a trunk link if the neighboring interface is set to **trunk** or **desirable**
`Switch(config-if)# switchport nonegotiate`	Prevents the interface from generating DTP frames **NOTE:** Use the **switchport mode nonegotiate** command only when the interface switchport mode is **access** or **trunk**. You must manually configure the neighboring interface to establish a trunk link
`Switch(config-if)# switchport mode trunk`	Puts the interface into permanent trunking mode and negotiates to convert the link into a trunk link **NOTE:** With the **switchport mode trunk** command set, the interface becomes a trunk link even if the neighboring interface is not a trunk link

TIP: The default mode is dependent on the platform. For the 2960/9200 series, the default mode is dynamic auto.

TIP: On a 2960/9200 series switch, the default for all ports is to be an access port. However, with the default DTP mode being dynamic auto, an access port can be converted into a trunk port if that port receives DTP information from the other side of the link and that other side is set to **trunk** or **desirable**. It is therefore recommended that you hard-code all access ports as access ports with the **switchport mode access** command. This way, DTP information will not inadvertently change an access port to a trunk port. Any port set with the **switchport mode access** command ignores any DTP requests to convert the link.

TIP: VLAN Trunking Protocol (VTP) domain names must match for a DTP to negotiate a trunk.

Setting the VLAN Encapsulation Type

Depending on the series of switch that you are using, you may have a choice as to what type of VLAN encapsulation you want to use: the Cisco proprietary Inter-Switch Link (ISL) or the IEEE Standard 802.1Q (dot1q). The 2960 and 9200 series of switches only support dot1q trunking.

CAUTION: Cisco ISL has been deprecated. Depending on the age and model of your Cisco switch, you may be able to change the encapsulation type between dot1q and ISL.

`Switch3650(config)# interface fastethernet 0/1`	Moves to interface configuration mode
`Switch3650(config-if)# switchport mode trunk`	Puts the interface into permanent trunking mode and negotiates to convert the link into a trunk link
`Switch3650(config-if)# switchport trunk encapsulation isl`	Specifies ISL encapsulation on the trunk link
`Switch3650(config-if)# switchport trunk encapsulation dot1q`	Specifies 802.1q tagging on the trunk link
`Switch3650(config-if)# switchport trunk encapsulation negotiate`	Specifies that the interface negotiate with the neighboring interface to become either an ISL or dot1q trunk, depending on the capabilities or configuration of the neighboring interface

TIP: With the **switchport trunk encapsulation negotiate** command set, the preferred trunking method is ISL.

CAUTION: The 2960, 2960-x, and 9200 series of switches support only dot1q trunking.

VLAN Trunking Protocol (VTP)

VTP is a Cisco proprietary protocol that allows for VLAN configuration (addition, deletion, or renaming of VLANs) to be consistently maintained across a common administrative domain.

`Switch(config)#` **`vtp mode client`**	Changes the switch to VTP client mode
`Switch(config)#` **`vtp mode server`**	Changes the switch to VTP server mode **NOTE:** By default, all Catalyst switches are in server mode
`Switch(config)#` **`vtp mode transparent`**	Changes the switch to VTP transparent mode
`Switch(config)#` **`no vtp mode`**	Returns the switch to the default VTP server mode
`Switch(config)#` **`vtp domain`** *`domain-name`*	Configures the VTP domain name. The name can be from 1 to 32 characters long **NOTE:** All switches operating in VTP server or client mode must have the same domain name to ensure communication
`Switch(config)#` **`vtp password`** *`password`*	Configures a VTP password. In Cisco IOS Software Release 12.3 and later, the password is an ASCII string from 1 to 32 characters long. If you are using a Cisco IOS Software release earlier than 12.3, the password length ranges from 8 to 64 characters long **NOTE:** To communicate with each other, all switches must have the same VTP password set
`Switch(config)#` **`vtp version`** *`number`*	Sets the VTP Version to Version 1, Version 2, or Version 3 **NOTE:** VTP versions are not interoperable. All switches must use the same version (with V1 and V2). The biggest difference between Versions 1 and 2 is that Version 2 has support for Token Ring VLANs. Version 3 has added new features such as the creation of a VTP primary server, to prevent the accidental deletion of VLANs that occurred in V1 and V2. V3 also supports extended VLANs, private VLANs, Multiple Spanning Tree Protocol (MSTP) and the ability to be disabled per interface as well as globally VTP Version 3 is compatible with Version 2, but not Version 1
`Switch2960#` **`vtp primary-server`**	Changes the operation state of a switch from a secondary server (the default state) to a primary server and advertises the configuration to the domain. If the switch password is configured as **hidden**, you are prompted to reenter the password. This happens only if configured in Version 2. This prompt occurs in privileged EXEC mode but not in global config mode
`Switch2960#` **`vtp primary-server vlan`**	(Optional) Selects the VLAN database as the takeover feature. This is the default state

Switch2960# **vtp primary-server mst**	(Optional) Selects the multiple spanning tree (MST) database as the takeover feature
Switch2960# **vtp primary-server force**	(Optional) Entering **force** overwrites the configuration of any conflicting servers. If you do not enter **force**, you are prompted for confirmation before the takeover
Switch9200# **vtp primary**	Configures the device as a VTP primary server on the 9200 series switch. The same optional parameters of **vlan**, **mst**, and **force** are also valid on this platform
Switch(config)# **vtp pruning**	Enables VTP pruning **NOTE:** By default, VTP pruning is disabled. You need to enable VTP pruning on only one switch in VTP server mode

NOTE: Only VLANs included in the pruning-eligible list can be pruned. VLANs 2 through 1001 are pruning eligible by default on trunk ports. Reserved VLANs and extended-range VLANs cannot be pruned. To change which eligible VLANs can be pruned, use the interface-specific **switchport trunk pruning vlan** command:

```
Switch(config-if)# switchport trunk pruning vlan remove 4,20-30
! Removes VLANs 4 and 20-30
Switch(config-if)# switchport trunk pruning vlan except 40-50
! All VLANs are added to the pruning list except for 40-50
```

CAUTION: Due to the inherent risk in having VTP servers overwrite each other and cause VLANs to disappear, Cisco recommends as a best practice deploying VTP in transparent mode. If you are going to use a client/server model, use Version 3 and the use of a VTPv3 primary server to prevent accidental database overwrites.

Verifying VTP

Switch# **show vtp status**	Displays general information about VTP configuration
Switch# **show vtp counters**	Displays the VTP counters for the switch

NOTE: If trunking has been established before VTP is set up, VTP information is propagated throughout the switch fabric almost immediately. However, because VTP information is advertised only every 300 seconds (5 minutes), unless a change has been made to force an update, it can take several minutes for VTP information to be propagated.

Inter-VLAN Communication Using an External Router: Router-on-a-Stick

`Router(config)#` **`interface`** **`fastethernet 0/0`**	Moves to interface configuration mode
`Router(config-if)#` **`duplex full`**	Sets the interface to full duplex
`Router(config-if)#` **`no shutdown`**	Enables the interface
`Router(config-if)#` **`interface`** **`fastethernet 0/0.1`**	Creates subinterface 0/0.1 and moves to subinterface configuration mode
`Router(config-subif)#` **`description Management VLAN 1`**	(Optional) Sets the locally significant description of the subinterface
`Router(config-subif)#` **`encapsulation dot1q 1 native`**	Assigns VLAN 1 to this subinterface. VLAN 1 will be the native VLAN. This subinterface uses the 802.1q tagging protocol
`Router(config-subif)#` **`ip address`** **`192.168.1.1 255.255.255.0`**	Assigns the IP address and netmask
`Router(config-subif)#` **`interface`** **`fastethernet 0/0.10`**	Creates subinterface 0/0.10 and moves to subinterface configuration mode
`Router(config-subif)#` **`description Accounting VLAN 10`**	(Optional) Sets the locally significant description of the subinterface
`Router(config-subif)#` **`encapsulation dot1q 10`**	Assigns VLAN 10 to this subinterface. This subinterface uses the 802.1q tagging protocol
`Router(config-subif)#` **`ip address`** **`192.168.10.1 255.255.255.0`**	Assigns the IP address and netmask
`Router(config-subif)#` **`exit`**	Returns to interface configuration mode
`Router(config-if)#` **`exit`**	Returns to global configuration mode
`Router(config)#`	

NOTE: The networks of the VLANs are directly connected to the router. Routing between these networks does not require a dynamic routing protocol. In a more complex topology, these routes need to either be advertised with whatever dynamic routing protocol is being used or be redistributed into whatever dynamic routing protocol is being used.

NOTE: Routes to the networks associated with these VLANs appear in the routing table as directly connected networks.

NOTE: In production environments, VLAN 1 should not be used as the management VLAN because it poses a potential security risk; all ports are in VLAN 1 by default, and it is an easy mistake to add a nonmanagement user to the management VLAN.

Inter-VLAN Communication on a Multilayer Switch Through a Switch Virtual Interface

NOTE: Rather than using an external router to provide inter-VLAN communication, a multilayer switch can perform the same task through the use of a switched virtual interface (SVI).

Removing L2 Switchport Capability of an Interface on an L3 Switch

Switch9300(config)# **interface gigabitethernet 0/1**	Moves to interface configuration mode
Switch9300(config-if)# **no switchport**	Creates a Layer 3 port on the switch **NOTE:** You can use the **no switchport** command on physical ports only on a Layer 3-capable switch

Configuring Inter-VLAN Communication on an L3 Switch

Switch9300(config)# **interface vlan 1**	Creates a virtual interface for VLAN 1 and enters interface configuration mode
Switch9300(config-if)# **ip address 172.16.1.1 255.255.255.0**	Assigns IP address and netmask
Switch9300(config-if)# **no shutdown**	Enables the interface
Switch9300(config)# **interface vlan 10**	Creates a virtual interface for VLAN 10 and enters interface configuration mode
Switch9300(config-if)# **ip address 172.16.10.1 255.255.255.0**	Assigns an IP address and netmask
Switch9300(config-if)# **no shutdown**	Enables the interface
Switch9300(config)# **interface vlan 20**	Creates a virtual interface for VLAN 20 and enters interface configuration mode
Switch9300(config-if)# **ip address 172.16.20.1 255.255.255.0**	Assigns an IP address and netmask
Switch9300(config-if)# **no shutdown**	Enables the interface
Switch9300(config-if)# **exit**	Returns to global configuration mode
Switch9300(config)# **ip routing**	Enables routing on the switch

NOTE: For an SVI to go to up/up and be added to the routing table, the VLAN for the SVI must be created, an IP address must be assigned, and at least one interface must support it.

Inter-VLAN Communication Tips

- Although most older routers (routers running IOS 12.2 and earlier) support both ISL and dot1q, some switch models support only dot1q, such as the 2960, 2960-x, and 9200 series. Check with the version of IOS you are using to determine whether ISL or dot1q is supported.
    - ISL will probably not be an option, as it has been deprecated for quite some time.

- If you need to use ISL as your trunking protocol, use the command **encapsulation isl** *x*, where *x* is the number of the VLAN to be assigned to that subinterface.

- Recommended best practice is to use the same number as the VLAN number for the subinterface number. It is easier to troubleshoot VLAN 10 on subinterface fa0/0.10 than on fa0/0.2

Configuration Example: Inter-VLAN Communication

Figure 10-1 illustrates the network topology for the configuration that follows, which shows how to configure inter-VLAN communication using commands covered in this chapter. Some commands used in this configuration are from other chapters.

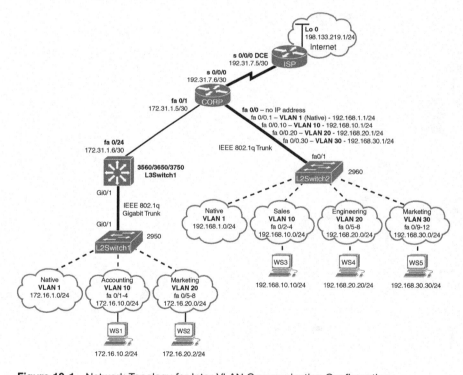

Figure 10-1   Network Topology for Inter-VLAN Communication Configuration

ISP Router

`Router> `**`enable`**	Moves to privileged EXEC mode
`Router># `**`configure terminal`**	Moves to global configuration mode
`Router(config)# `**`hostname ISP`**	Sets the host name
`ISP(config)# `**`interface loopback 0`**	Moves to interface configuration mode

`ISP(config-if)# description simulated address representing remote website`	Sets the locally significant interface description
`ISP(config-if)# ip address 198.133.219.1 255.255.255.0`	Assigns an IP address and netmask
`ISP(config-if)# interface serial 0/0/0`	Moves to interface configuration mode
`ISP(config-if)# description WAN link to the Corporate Router`	Sets the locally significant interface description
`ISP(config-if)# ip address 192.31.7.5 255.255.255.252`	Assigns an IP address and netmask
`ISP(config-if)# clock rate 56000`	Assigns a clock rate to the interface; DCE cable is plugged into this interface
`ISP(config-if)# no shutdown`	Enables the interface
`ISP(config-if)# exit`	Returns to global configuration mode
`ISP(config-if)# router eigrp 10`	Creates Enhanced Interior Gateway Routing Protocol (EIGRP) routing process 10
`ISP(config-router)# network 198.133.219.0`	Advertises directly connected networks (classful address only)
`ISP(config-router)# network 192.31.7.0`	Advertises directly connected networks (classful address only)
`ISP(config-router)# no auto-summary`	Disables automatic summarization
`ISP(config-router)# exit`	Returns to global configuration mode
`ISP(config)# exit`	Returns to privileged EXEC mode
`ISP# copy running-config startup-config`	Saves the configuration to NVRAM

CORP Router

`Router> enable`	Moves to privileged EXEC mode
`Router># configure terminal`	Moves to global configuration mode
`Router(config)# hostname CORP`	Sets the host name
`CORP(config)# no ip domain-lookup`	Turns off Domain Name System (DNS) resolution to avoid wait time due to DNS lookup of spelling errors
`CORP(config)# interface serial 0/0/0`	Moves to interface configuration mode
`CORP(config-if)# description link to ISP`	Sets the locally significant interface description
`CORP(config-if)# ip address 192.31.7.6 255.255.255.252`	Assigns an IP address and netmask
`CORP(config-if)# no shutdown`	Enables the interface

`CORP(config)# interface fastethernet 0/1`	Moves to interface configuration mode
`CORP(config-if)# description link to L3Switch1`	Sets the locally significant interface description
`CORP(config-if)# ip address 172.31.1.5 255.255.255.252`	Assigns an IP address and netmask
`CORP(config-if)# no shutdown`	Enables the interface
`CORP(config-if)# exit`	Returns to global configuration mode
`CORP(config)# interface fastethernet 0/0`	Enters interface configuration mode
`CORP(config-if)# duplex full`	Enables full-duplex operation to ensure trunking takes effect between here and L2Switch2
`CORP(config-if)# no shutdown`	Enables the interface
`CORP(config-if)# interface fastethernet 0/0.1`	Creates a virtual subinterface and moves to subinterface configuration mode
`CORP(config-subif)# description Management VLAN 1 - Native VLAN`	Sets the locally significant interface description
`CORP(config-subif)# encapsulation dot1q 1 native`	Assigns VLAN 1 to this subinterface. VLAN 1 is the native VLAN. This subinterface uses the 802.1q protocol
`CORP(config-subif)# ip address 192.168.1.1 255.255.255.0`	Assigns an IP address and netmask
`CORP(config-subif)# interface fastethernet 0/0.10`	Creates a virtual subinterface and moves to subinterface configuration mode
`CORP(config-subif)# description Sales VLAN 10`	Sets the locally significant interface description
`CORP(config-subif)# encapsulation dot1q 10`	Assigns VLAN 10 to this subinterface. This subinterface uses the 802.1q protocol
`CORP(config-subif)# ip address 192.168.10.1 255.255.255.0`	Assigns an IP address and netmask
`CORP(config-subif)# interface fastethernet 0/0.20`	Creates a virtual subinterface and moves to subinterface configuration mode
`CORP(config-subif)# description Engineering VLAN 20`	Sets the locally significant interface description
`CORP(config-subif)# encapsulation dot1q 20`	Assigns VLAN 20 to this subinterface. This subinterface uses the 802.1q protocol
`CORP(config-subif)# ip address 192.168.20.1 255.255.255.0`	Assigns an IP address and netmask
`CORP(config-subif)# interface fastethernet 0/0.30`	Creates a virtual subinterface and moves to subinterface configuration mode
`CORP(config-subif)# description Marketing VLAN 30`	Sets the locally significant interface description

CORP(config-subif)# **encapsulation dot1q 30**	Assigns VLAN 30 to this subinterface. This subinterface uses the 802.1q protocol
CORP(config-subif)# **ip add 192.168.30.1 255.255.255.0**	Assigns an IP address and netmask
CORP(config-subif)# **exit**	Returns to interface configuration mode
CORP(config-if)# **exit**	Returns to global configuration mode
CORP(config)# **router eigrp 10**	Creates EIGRP routing process 10 and moves to router configuration mode
CORP(config-router)# **network 192.168.1.0**	Advertises the 192.168.1.0 network
CORP(config-router)# **network 192.168.10.0**	Advertises the 192.168.10.0 network
CORP(config-router)# **network 192.168.20.0**	Advertises the 192.168.20.0 network
CORP(config-router)# **network 192.168.30.0**	Advertises the 192.168.30.0 network
CORP(config-router)# **network 172.31.0.0**	Advertises the 172.31.0.0 network
CORP(config-router)# **network 192.31.7.0**	Advertises the 192.31.7.0 network
CORP(config-router)# **no auto-summary**	Turns off automatic summarization at classful boundary
CORP(config-router)# **exit**	Returns to global configuration mode
CORP(config)# **exit**	Returns to privileged EXEC mode
CORP# **copy running-config startup-config**	Saves the configuration in NVRAM

L2Switch2 (Catalyst 2960)

Switch> **enable**	Moves to privileged EXEC mode
Switch# **configure terminal**	Moves to global configuration mode
Switch(config)# **hostname L2Switch2**	Sets the host name
L2Switch2(config)# **no ip domain-lookup**	Turns off DNS resolution
L2Switch2(config)# **vlan 10**	Creates VLAN 10 and enters VLAN configuration mode
L2Switch2(config-vlan)# **name Sales**	Assigns a name to the VLAN
L2Switch2(config-vlan)# **exit**	Returns to global configuration mode
L2Switch2(config)# **vlan 20**	Creates VLAN 20 and enters VLAN configuration mode

`L2Switch2(config-vlan)# name Engineering`	Assigns a name to the VLAN
`L2Switch2(config-vlan)# vlan 30`	Creates VLAN 30 and enters VLAN configuration mode. **NOTE:** You do not have to exit back to global configuration mode to execute this command
`L2Switch2(config-vlan)# name Marketing`	Assigns a name to the VLAN
`L2Switch2(config-vlan)# exit`	Returns to global configuration mode
`L2Switch2(config)# interface range fastethernet 0/2 - 4`	Enables you to set the same configuration parameters on multiple ports at the same time
`L2Switch2(config-if-range)# switchport mode access`	Sets ports 2 to 4 as access ports
`L2Switch2(config-if-range)# switchport access vlan 10`	Assigns ports 2 to 4 to VLAN 10
`L2Switch2(config-if-range)# interface range fastethernet 0/5 - 8`	Enables you to set the same configuration parameters on multiple ports at the same time
`L2Switch2(config-if-range)# switchport mode access`	Sets ports 5 to 8 as access ports
`L2Switch2(config-if-range)# switchport access vlan 20`	Assigns ports 5 to 8 to VLAN 20
`L2Switch2(config-if-range)# interface range fastethernet 0/9 - 12`	Enables you to set the same configuration parameters on multiple ports at the same time
`L2Switch2(config-if-range)# switchport mode access`	Sets ports 9 to 12 as access ports
`L2Switch2(config-if-range)# switchport access vlan 30`	Assigns ports 9 to 12 to VLAN 30
`L2Switch2(config-if-range)# exit`	Returns to global configuration mode
`L2Switch2(config)# interface fastethernet 0/1`	Moves to interface configuration mode
`L2Switch2(config)# description Trunk Link to CORP Router`	Sets the locally significant interface description
`L2Switch2(config-if)# switchport mode trunk`	Puts the interface into trunking mode and negotiates to convert the link into a trunk link
`L2Switch2(config-if)# exit`	Returns to global configuration mode
`L2Switch2(config)# interface vlan 1`	Creates a virtual interface for VLAN 1 and enters interface configuration mode
`L2Switch2(config-if)# ip address 192.168.1.2 255.255.255.0`	Assigns an IP address and netmask
`L2Switch2(config-if)# no shutdown`	Enables the interface

`L2Switch2(config-if)# exit`	Returns to global configuration mode
`L2Switch2(config)# ip default-gateway 192.168.1.1`	Assigns a default gateway address
`L2Switch2(config)# exit`	Returns to privileged EXEC mode
`L2Switch2# copy running-config startup-config`	Saves the configuration in NVRAM

L3Switch1 (Catalyst 3560/3650/3750)

`Switch> enable`	Moves to privileged EXEC mode
`Switch# configure terminal`	Moves to global configuration mode
`Switch(config)# hostname L3Switch1`	Sets the host name
`L3Switch1(config)# no ip domain-lookup`	Turns off DNS queries so that spelling mistakes do not slow you down
`L3Switch1(config)# vtp mode server`	Changes the switch to VTP server mode
`L3Switch1(config)# vtp domain testdomain`	Configures the VTP domain name to *testdomain*
`L3Switch1(config)# vlan 10`	Creates VLAN 10 and enters VLAN configuration mode
`L3Switch1(config-vlan)# name Accounting`	Assigns a name to the VLAN
`L3Switch1(config-vlan)# exit`	Returns to global configuration mode
`L3Switch1(config)# vlan 20`	Creates VLAN 20 and enters VLAN configuration mode
`L3Switch1(config-vlan)# name Marketing`	Assigns a name to the VLAN
`L3Switch1(config-vlan)# exit`	Returns to global configuration mode
`L3Switch1(config)# interface gigabitethernet 0/1`	Moves to interface configuration mode
`L3Switch1(config-if)# switchport trunk encapsulation dot1q`	Specifies 802.1q tagging on the trunk link
`L3Switch1(config-if)# switchport mode trunk`	Puts the interface into trunking mode and negotiates to convert the link into a trunk link
`L3Switch1(config-if)# exit`	Returns to global configuration mode
`L3Switch1(config)# ip routing`	Enables IP routing on this device
`L3Switch1(config)# interface vlan 1`	Creates a virtual interface for VLAN 1 and enters interface configuration mode
`L3Switch1(config-if)# ip address 172.16.1.1 255.255.255.0`	Assigns an IP address and netmask
`L3Switch1(config-if)# no shutdown`	Enables the interface

`L3Switch1(config-if)# interface vlan 10`	Creates a virtual interface for VLAN 10 and enters interface configuration mode
`L3Switch1(config-if)# ip address 172.16.10.1 255.255.255.0`	Assigns an IP address and mask
`L3Switch1(config-if)# no shutdown`	Enables the interface
`L3Switch1(config-if)# interface vlan 20`	Creates a virtual interface for VLAN 20 and enters interface configuration mode
`L3Switch1(config-if)# ip address 172.16.20.1 255.255.255.0`	Assigns an IP address and mask
`L3Switch1(config-if)# no shutdown`	Enables the interface
`L3Switch1(config-if)# exit`	Returns to global configuration mode
`L3Switch1(config)# interface fastethernet 0/24`	Enters interface configuration mode
`L3Switch1(config-if)# no switchport`	Creates a Layer 3 port on the switch
`L3Switch1(config-if)# ip address 172.31.1.6 255.255.255.252`	Assigns an IP address and netmask
`L3Switch1(config-if)# exit`	Returns to global configuration mode
`L3Switch1(config)# router eigrp 10`	Creates EIGRP routing process 10 and moves to router configuration mode
`L3Switch1(config-router)# network 172.16.0.0`	Advertises the 172.16.0.0 classful network
`L3Switch1(config-router)# network 172.31.0.0`	Advertises the 172.31.0.0 classful network
`L3Switch1(config-router)# no auto-summary`	Turns off automatic summarization at classful boundary
`L3Switch1(config-router)# exit`	Applies changes and returns to global configuration mode
`L3Switch1(config)# exit`	Returns to privileged EXEC mode
`L3Switch1# copy running-config startup-config`	Saves configuration in NVRAM

L2Switch1 (Catalyst 2960)

`Switch> enable`	Moves to privileged EXEC mode
`Switch# configure terminal`	Moves to global configuration mode
`Switch(config)# hostname L2Switch1`	Sets the host name
`L2Switch1(config)# no ip domain-lookup`	Turns off DNS queries so that spelling mistakes do not slow you down
`L2Switch1(config)# vtp domain testdomain`	Configures the VTP domain name to *testdomain*
`L2Switch1(config)# vtp mode client`	Changes the switch to VTP client mode

`L2Switch1(config)# interface range fastethernet 0/1 - 4`	Enables you to set the same configuration parameters on multiple ports at the same time
`L2Switch1(config-if-range)# switchport mode access`	Sets ports 1 to 4 as access ports
`L2Switch1(config-if-range)# switchport access vlan 10`	Assigns ports 1 to 4 to VLAN 10
`L2Switch1(config-if-range)# interface range fastethernet 0/5 - 8`	Enables you to set the same configuration parameters on multiple ports at the same time
`L2Switch1(config-if-range)# switchport mode access`	Sets ports 5 to 8 as access ports
`L2Switch1(config-if-range)# switchport access vlan 20`	Assigns ports 5 to 8 to VLAN 20
`L2Switch1(config-if-range)# exit`	Returns to global configuration mode
`L2Switch1(config)# interface gigabitethernet 0/1`	Moves to interface configuration mode
`L2Switch1(config-if)# switchport mode trunk`	Puts the interface into trunking mode and negotiates to convert the link into a trunk link
`L2Switch1(config-if)# exit`	Returns to global configuration mode
`L2Switch1(config)# interface vlan 1`	Creates a virtual interface for VLAN 1 and enters interface configuration mode
`L2Switch1(config-if)# ip address 172.16.1.2 255.255.255.0`	Assigns an IP address and netmask
`L2Switch1(config-if)# no shutdown`	Enables the interface
`L2Switch1(config-if)# exit`	Returns to global configuration mode
`L2Switch1(config)# ip default-gateway 172.16.1.1`	Assigns the default gateway address
`L2Switch1(config)# exit`	Returns to privileged EXEC mode
`L2Switch1# copy running-config startup-config`	Saves the configuration in NVRAM

CHAPTER 11

Spanning Tree Protocol

This chapter provides information and commands concerning the following topics:

- Spanning Tree Protocol definition
- Enabling Spanning Tree Protocol
- Changing the spanning-tree mode
- Configuring the root switch
- Configuring a secondary root switch
- Configuring port priority
- Configuring the path cost
- Configuring the switch priority of a VLAN
- Configuring STP timers
- Configuring optional spanning-tree features
    - PortFast
    - BPDU Guard (2xxx/3xxx series)
    - BPDU Guard (9xxx series)
- Enabling the extended system ID
- Verifying STP
- Troubleshooting Spanning Tree Protocol
- Configuration example: PVST+

Spanning Tree Protocol Definition

The spanning tree standards offer the same safety that routing protocols provide in Layer 3 forwarding environments to Layer 2 bridging environments. A single best path to a main bridge is found and maintained in the Layer 2 domain, and other redundant paths are managed by selective port blocking. Appropriate blocked ports begin forwarding when primary paths to the main bridge are no longer available.

There are several different spanning-tree modes and protocols:

- **Per VLAN Spanning Tree (PVST+):** This spanning-tree mode is based on the IEEE 802.1D standard and Cisco proprietary extensions. The PVST+ runs on each VLAN on the device up to the maximum supported, ensuring that each has a loop-free path through the network. PVST+ provides Layer 2 load balancing for the VLAN on which it runs. You can create different logical topologies by using the VLANs on your network to ensure that all of your links are used but that no one link is oversubscribed. Each instance of PVST+ on a VLAN has a single root device. This root device propagates the spanning-tree information associated with that VLAN to all other devices in the network. Because each device has the same information about the network, this process ensures that the network topology is maintained.

- **Rapid PVST+:** This spanning-tree mode is the same as PVST+ except that it uses a rapid convergence based on the IEEE 802.1w standard. Beginning from Cisco IOS Release 15.2(4)E, the STP default mode is Rapid PVST+. To provide rapid convergence, Rapid PVST+ immediately deletes dynamically learned MAC address entries on a per-port basis upon receiving a topology change. By contrast, PVST+ uses a short aging time for dynamically learned MAC address entries. Rapid PVST+ uses the same configuration as PVST+ and the device needs only minimal extra configuration. The benefit of Rapid PVST+ is that you can migrate a large PVST+ install base to Rapid PVST+ without having to learn the complexities of the Multiple Spanning Tree Protocol (MSTP) configuration and without having to reprovision your network. In Rapid PVST+ mode, each VLAN runs its own spanning-tree instance up to the maximum supported.

- **Multiple Spanning Tree Protocol (MSTP):** This spanning-tree mode is based on the IEEE 802.1s standard. You can map multiple VLANs to the same spanning-tree instance, which reduces the number of spanning-tree instances required to support a large number of VLANs. MSTP runs on top of the Rapid Spanning Tree Protocol (RSTP) (based on IEEE 802.1w), which provides for rapid convergence of the spanning tree by eliminating the forward delay and by quickly transitioning root ports and designated ports to the forwarding state. In a device stack, the cross-stack rapid transition (CSRT) feature performs the same function as RSTP. You cannot run MSTP without RSTP or CSRT.

NOTE: Default spanning-tree implementation for Catalyst 2950, 2960, 3550, 3560, and 3750 switches is PVST+. This is a per-VLAN implementation of 802.1D. Beginning from Cisco IOS Release 15.2(4)E, the STP default mode is Rapid PVST+ on all switch platforms.

Enabling Spanning Tree Protocol

`Switch(config)# ` **`spanning-tree vlan 5`**	Enables STP on VLAN 5
`Switch(config)# ` **`no spanning-tree vlan 5`**	Disables STP on VLAN 5

NOTE: Many access switches such as the Catalyst 2960, 3550, 3560, 3750, 9200, and 9300 support a maximum 128 spanning trees using any combination of PVST+ or Rapid PVST+. The 2950 model supports only 64 instances. Any VLANs created in excess of 128 spanning trees cannot have a spanning-tree instance running in them. There is a possibility of an L2 loop that could not be broken in the case where a VLAN without spanning tree is transported across a trunk. It is recommended that you use MSTP if the number of VLANs in a common topology is high.

CAUTION: Spanning tree is enabled by default on VLAN 1 and on all newly created VLANs up to the spanning-tree limit. Disable spanning tree only if you are sure there are no loops in the network topology. When spanning tree is disabled and loops are present in the topology, excessive traffic and indefinite packet duplication can drastically reduce network performance. Networks have been known to crash in seconds due to broadcast storms created by loops.

Changing the Spanning-Tree Mode

You can configure different types of spanning trees on a Cisco switch. The options vary according to the platform.

`Switch(config)#` **`spanning-tree mode pvst`**	Enables PVST+. This is the default setting
`Switch(config)#` **`spanning-tree mode mst`**	Enters MST mode
`Switch(config)#` **`spanning-tree mst configuration`**	Enters MST configuration submode **NOTE:** Use the command **no spanning-tree mst configuration** to clear the MST configuration
`Switch(config)#` **`spanning-tree mode rapid-pvst`**	Enables Rapid PVST+
`Switch#` **`clear spanning-tree detected-protocols`**	If any port on the device is connected to a port on a legacy IEEE 802.1D device, this command restarts the protocol migration process on the entire device This step is optional if the designated device detects that this device is running Rapid PVST+

BPDU Guard (3650/9xxx Series)

You can enable the BPDU Guard feature if your switch is running PVST+, Rapid PVST+, or MSTP.

The Bridge Protocol Data Unit (BPDU) Guard feature can be globally enabled on the switch or can be enabled per port.

When you enable BPDU Guard at the global level on PortFast-enabled ports, spanning tree shuts down ports that are in a PortFast-operational state if any BPDU is received on them. When you enable BPDU Guard at the interface level on any port without also enabling the PortFast feature, and the port receives a BPDU, it is put in the error-disabled state.

`Switch(config)#` **`spanning-tree portfast bpduguard default`**	Enables BPDU guard globally **NOTE:** By default, BPDU Guard is disabled.
`Switch(config)#` **`interface gigabitethernet 1/0/2`**	Enters into interface configuration mode
`Switch(config-if)#` **`spanning-tree portfast edge`**	Enables the PortFast edge feature
`Switch(config-if)#` **`end`**	Returns to privileged EXEC mode

Configuring the Root Switch

`Switch(config)# spanning-tree vlan 5 root {primary	secondary}`	Modifies the switch priority from the default 32768 to a lower value to allow the switch to become the primary or secondary root switch for VLAN 5 (depending on which argument is chosen) **NOTE:** This switch sets its priority to 24576. If any other switch has a priority set to below 24576 already, this switch sets its own priority to 4096 *less* than the lowest switch priority. If by doing this the switch has a priority of less than 1, this command fails
`Switch(config)# spanning-tree vlan 5 root primary`	Configures the switch to become the root switch for VLAN 5 **NOTE:** The maximum switch topology width and the hello-time can be set within this command **TIP:** The root switch should be a backbone or distribution switch
`Switch(config)# spanning-tree vlan 5 root primary diameter 7`	Configures the switch to be the root switch for VLAN 5 and sets the network diameter to 7 **TIP:** The **diameter** keyword defines the maximum number of switches between any two end stations. The range is from 2 to 7 switches **TIP:** The **hello-time** keyword sets the hello-interval timer to any amount between 1 and 10 seconds. The default time is 2 seconds

Configuring a Secondary Root Switch

`Switch(config)# spanning-tree vlan 5 root secondary`	Configures the switch to become the root switch for VLAN 5 should the primary root switch fail **NOTE:** This switch resets its priority to 28672. If the root switch fails and all other switches are set to the default priority of 32768, this becomes the new root switch
`Switch(config)# spanning-tree vlan 5 root secondary diameter 7`	Configures the switch to be the secondary root switch for VLAN 5 and sets the network diameter to 7

Configuring Port Priority

`Switch(config)# interface gigabitethernet 0/1`	Moves to interface configuration mode
`Switch(config-if)# spanning-tree port-priority 64`	Configures the port priority for the interface that is an access port

`Switch(config-if)#` **`spanning-tree vlan 5`** **`port-priority 64`**	Configures the VLAN port priority for an interface that is a trunk port **NOTE:** If a loop occurs, spanning tree uses the port priority when selecting an interface to put into the forwarding state. Assign a higher priority value (lower numerical number) to interfaces you want selected first and a lower priority value (higher numerical number) to interfaces you want selected last The number can be between 0 and 240 in increments of 16. The default port priority is 128

NOTE: The **port priority** setting supersedes the physical port number in spanning tree calculations.

Configuring the Path Cost

`Switch(config)#` **`interface`** **`gigabitethernet 0/1`**	Moves to interface configuration mode
`Switch(config-if)#` **`spanning-tree cost 100000`**	Configures the cost for the interface that is an access port. The range is 1 to 200000000; the default value is derived from the media speed of the interface
`Switch(config-if)#` **`spanning-tree vlan 5`** **`cost 1500000`**	Configures the VLAN cost for an interface that is a trunk port. The VLAN number can be specified as a single VLAN ID number, a range of VLANs separated by a hyphen, or a series of VLANs separated by a comma. The range is 1 to 4094. For the cost, the range is 1 to 200000000; the default value is derived from the media speed of the interface **NOTE:** If a loop occurs, STP uses the path cost when trying to determine which interface to place into the forwarding state. A higher path cost means a lower-speed transmission

Configuring the Switch Priority of a VLAN

`Switch(config)#` **`spanning-`** **`tree vlan 5 priority 12288`**	Configures the switch priority of VLAN 5 to 12288

NOTE: With the **priority** keyword, the range is 0 to 61440 in increments of 4096. The default is 32768. The lower the priority, the more likely the switch will be chosen as the root switch. Only the following numbers can be used as priority values:

0	4096	8192	12288
16384	20480	24576	28672
32768	36864	40960	45056
49152	53248	57344	61440

CAUTION: Cisco recommends caution when using this command. Cisco further recommends that the **spanning-tree vlan** *x* **root primary** or the **spanning-tree vlan** *x* **root secondary** command be used instead to modify the switch priority.

Configuring STP Timers

`Switch(config)#` **spanning-tree vlan 5 hello-time 4**	Changes the hello-delay timer to 4 seconds on VLAN 5
`Switch(config)#` **spanning-tree vlan 5 forward-time 20**	Changes the forward-delay timer to 20 seconds on VLAN 5
`Switch(config)#` **spanning-tree vlan 5 max-age 25**	Changes the maximum-aging timer to 25 seconds on VLAN 5

NOTE: For the **hello-time** command, the range is 1 to 10 seconds. The default is 2 seconds.

For the **forward-time** command, the range is 4 to 30 seconds. The default is 15 seconds.

For the **max-age** command, the range is 6 to 40 seconds. The default is 20 seconds.

Configuring Optional Spanning-Tree Features

Although the following commands are not mandatory for STP to work, you might find these helpful to fine-tune your network.

PortFast

`Switch(config)#` **interface fastethernet 0/10**	Moves to interface configuration mode
`Switch(config-if)#` **spanning-tree portfast**	Enables PortFast on an access port
`Switch(config-if)#` **spanning-tree portfast trunk**	Enables PortFast on a trunk port **CAUTION:** Use the PortFast command only when connecting a single end station to an access or trunk port. Using this command on a port connected to a switch or hub might prevent spanning tree from detecting loops **NOTE:** If you enable the voice VLAN feature, PortFast is enabled automatically. If you disable voice VLAN, PortFast is still enabled
`Switch(config)#` **spanning-tree portfast default**	Globally enables PortFast on all switchports that are nontrunking **NOTE:** You can override the **spanning-tree portfast default** global configuration command by using the **spanning-tree portfast disable** interface configuration command

Switch# `show spanning-tree interface fastethernet 0/10 portfast`	Displays PortFast information on interface fastethernet 0/10

BPDU Guard (2xxx/Older 3xxx Series)

Switch(config)# `spanning-tree portfast bpduguard default`	Globally enables BPDU Guard on ports where **portfast** is enabled
Switch(config)# `interface range fastethernet 0/1 - 5`	Enters interface range configuration mode
Switch(config-if-range)# `spanning-tree portfast`	Enables PortFast on all interfaces in the range **NOTE:** Best practice is to enable PortFast at the same time as BPDU Guard
Switch(config-if-range)# `spanning-tree bpduguard enable`	Enables BPDU Guard on the interface **NOTE:** By default, BPDU Guard is disabled
Switch(config-if)# `spanning-tree bpduguard disable`	Disables BPDU Guard on the interface
Switch(config)# `errdisable recovery cause bpduguard`	Allows port to reenable itself if the cause of the error is BPDU Guard by setting a recovery timer
Switch(config)# `errdisable recovery interval 400`	Sets recovery timer to 400 seconds. The default is 300 seconds. The range is from 30 to 86,400 seconds
Switch# `show spanning-tree summary totals`	Verifies whether BPDU Guard is enabled or disabled
Switch# `show errdisable recovery`	Displays errdisable recovery timer information

Enabling the Extended System ID

Switch(config)# `spanning-tree extend system-id`	Enables the extended system ID, also known as MAC address reduction **NOTE:** Catalyst switches running software earlier than Cisco IOS Release 12.1(8) EA1 do not support the extended system ID
Switch# `show spanning-tree summary`	Verifies that the extended system ID is enabled
Switch# `show spanning-tree bridge`	Displays the extended system ID as part of the bridge ID **NOTE:** The 12-bit extended system ID is the VLAN number for the instance of PVST+ and PVRST+ spanning tree. In MST, these 12 bits carry the instance number

Verifying STP

`Switch# show spanning-tree`	Displays STP information
`Switch# show spanning-tree active`	Displays STP information on active interfaces only
`Switch# show spanning-tree bridge`	Displays status and configuration of this bridge
`Switch# show spanning-tree detail`	Displays a detailed summary of interface information
`Switch# show spanning-tree interface gigabitethernet 1/0/1`	Displays STP information for interface gigabitethernet 1/0/1
`Switch# show spanning-tree summary`	Displays a summary of port states
`Switch# show spanning-tree summary totals`	Displays the total lines of the STP section
`Switch# show spanning-tree vlan 5`	Displays STP information for VLAN 5

Troubleshooting Spanning Tree Protocol

`Switch# debug spanning-tree all`	Displays all spanning-tree debugging events
`Switch# debug spanning-tree events`	Displays spanning-tree debugging topology events
`Switch# debug spanning-tree backbonefast`	Displays spanning-tree debugging BackboneFast events
`Switch# debug spanning-tree uplinkfast`	Displays spanning-tree debugging UplinkFast events
`Switch# debug spanning-tree mstp all`	Displays all MST debugging events
`Switch# debug spanning-tree switch state`	Displays spanning-tree port state changes
`Switch# debug spanning-tree pvst+`	Displays PVST+ events

Configuration Example: PVST+

Figure 11-1 shows the network topology for the configuration of PVST+ using commands covered in this chapter. Assume that other commands needed for connectivity have already been configured.

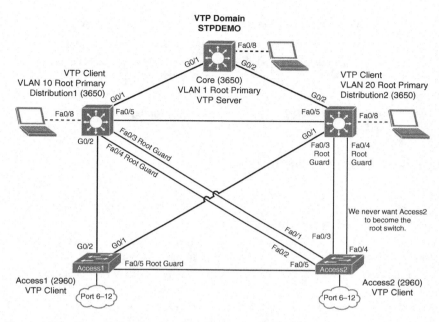

Figure 11-1 Network Topology for STP Configuration Example

Core Switch (3650)

`Switch> ` **`enable`**	Moves to privileged EXEC mode
`Switch# ` **`configure terminal`**	Moves to global configuration mode
`Switch(config)# ` **`hostname Core`**	Sets host name
`Core(config)# ` **`no ip domain-lookup`**	Turns off Dynamic Name System (DNS) queries so that spelling mistakes do not slow you down
`Core(config)# ` **`vtp mode server`**	Changes the switch to VTP server mode. This is the default mode
`Core(config)# ` **`vtp domain STPDEMO`**	Configures the VTP domain name to *STPDEMO*
`Core(config)# ` **`vlan 10`**	Creates VLAN 10 and enters VLAN configuration mode
`Core(config-vlan)# ` **`name Accounting`**	Assigns a name to the VLAN
`Core(config-vlan)# ` **`exit`**	Returns to global configuration mode
`Core(config)# ` **`vlan 20`**	Creates VLAN 20 and enters VLAN configuration mode

Core(config-vlan)# **name Marketing**	Assigns a name to the VLAN
Core(config-vlan)# **exit**	Returns to global configuration mode
Core(config)# **spanning-tree vlan 1 root primary**	Configures the switch to become the root switch for VLAN 1
Core(config)# **exit**	Returns to privileged EXEC mode
Core# **copy running-config startup-config**	Saves the configuration to NVRAM

Distribution 1 Switch (3650)

Switch> **enable**	Moves to privileged EXEC mode
Switch# **configure terminal**	Moves to global configuration mode
Switch(config)# **hostname Distribution1**	Sets host name
Distribution1(config)# **no ip domain-lookup**	Turns off DNS queries so that spelling mistakes do not slow you down
Distribution1(config)# **vtp domain STPDEMO**	Configures the VTP domain name to *STPDEMO*
Distribution1(config)# **vtp mode client**	Changes the switch to VTP client mode
Distribution1(config)# **spanning-tree vlan 10 root primary**	Configures the switch to become the root switch of VLAN 10
Distribution1(config)# **exit**	Returns to privileged EXEC mode
Distribution1# **copy running-config startup-config**	Saves the configuration to NVRAM

Distribution 2 Switch (3650)

Switch> **enable**	Moves to privileged EXEC mode
Switch# **configure terminal**	Moves to global configuration mode
Switch(config)# **hostname Distribution2**	Sets the host name
Distribution2(config)# **no ip domain-lookup**	Turns off DNS queries so that spelling mistakes do not slow you down
Distribution2(config)# **vtp domain STPDEMO**	Configures the VTP domain name to *STPDEMO*
Distribution2(config)# **vtp mode client**	Changes the switch to VTP client mode
Distribution2(config)# **spanning-tree vlan 20 root primary**	Configures the switch to become the root switch of VLAN 20
Distribution2(config)# **exit**	Returns to privileged EXEC mode
Distribution2# **copy running-config startup-config**	Saves the configuration to NVRAM

Access 1 Switch (2960)

`Switch> enable`	Moves to privileged EXEC mode
`Switch# configure terminal`	Moves to global configuration mode
`Switch(config)# hostname Access1`	Sets the host name
`Access1(config)# no ip domain-lookup`	Turns off DNS queries so that spelling mistakes do not slow you down
`Access1(config)# vtp domain STPDEMO`	Configures the VTP domain name to *STPDEMO*
`Access1(config)# vtp mode client`	Changes the switch to VTP client mode
`Access1(config)# interface range fastethernet 0/6 - 12`	Moves to interface range configuration mode
`Access1(config-if-range)# switchport mode access`	Places all interfaces in switchport access mode
`Access1(config-if-range)# spanning-tree portfast`	Places all ports directly into forwarding mode
`Access1(config-if-range)# spanning-tree bpduguard enable`	Enables BPDU Guard
`Access1(config-if-range)# exit`	Moves back to global configuration mode
`Access1(config)# exit`	Returns to privileged EXEC mode
`Access1# copy running-config startup-config`	Saves the configuration to NVRAM

Access 2 Switch (2960)

`Switch> enable`	Moves to privileged EXEC mode
`Switch# configure terminal`	Moves to global configuration mode
`Switch(config)# hostname Access2`	Sets the host name
`Access2(config)# no ip domain-lookup`	Turns off DNS queries so that spelling mistakes do not slow you down
`Access2(config)# vtp domain STPDEMO`	Configures the VTP domain name to *STPDEMO*
`Access2(config)# vtp mode client`	Changes the switch to VTP client mode
`Access2(config)# interface range fastethernet 0/6 - 12`	Moves to interface range configuration mode
`Access2(config-if-range)# switchport mode access`	Places all interfaces in switchport access mode
`Access2(config-if-range)# spanning-tree portfast`	Places all ports directly into forwarding mode
`Access2(config-if-range)# spanning-tree bpduguard enable`	Enables BPDU Guard

`Access2(config-if-range)# exit`	Moves back to global configuration mode
`Access2(config)# spanning-tree vlan 1,10,20 priority 61440`	Ensures this switch does not become the root switch for VLAN 10
`Access2(config)# exit`	Returns to privileged EXEC mode
`Access2# copy running-config startup-config`	Saves config to NVRAM

Spanning-Tree Migration Example: PVST+ to Rapid-PVST+

The topology in Figure 11-1 is used for this migration example and adds to the configuration of the previous example.

Rapid-PVST+ uses the same BPDU format as the 802.1D. This interoperability between the two spanning tree protocols enables a longer conversion time in large networks without disrupting services.

The Spanning Tree features UplinkFast and BackboneFast in 802.1D-based PVST+ are already incorporated in the 802.1w-based Rapid-PVST+ and are disabled when you enable Rapid-PVST+. The 802.1D-based features of PVST+ such as PortFast, BPDU Guard, BPDU filter, root guard, and loop guard are applicable in Rapid-PVST+ mode and need not be changed.

NOTE: The 802.1D-based features of PVST+ are not part of the CCNA 200-301 exam topics; they are, however, part of the CCNP Implementing Cisco Enterprise Network Core Technologies (ENCOR 300-401) exam topics.

Access 1 Switch (2960)

`Access1> enable`	Moves to privileged EXEC mode
`Access1# configure terminal`	Moves to global configuration mode
`Access1(config)# spanning-tree mode rapid-pvst`	Enables 802.1w-based Rapid-PVST+
`Access1(config)# no spanning-tree uplinkfast`	Removes UplinkFast programming line
`Access1(config)# no spanning-tree backbonefast`	Removes BackboneFast programming line

Access 2 Switch (2960)

`Access2> enable`	Moves to privileged EXEC mode
`Access2# configure terminal`	Moves to global configuration mode
`Access2 (config)# spanning-tree mode rapid-pvst`	Enables 802.1w-based Rapid-PVST+

Distribution 1 Switch (3650)

`Distribution1> ` **`enable`**	Moves to privileged EXEC mode
`Distribution1# ` **`configure terminal`**	Moves to global configuration mode
`Distribution1 (config)# ` **`spanning-tree mode rapid-pvst`**	Enables 802.1w-based Rapid-PVST+

Distribution 2 Switch (3650)

`Distribution2> ` **`enable`**	Moves to privileged EXEC mode
`Distribution2# ` **`configure terminal`**	Moves to global configuration mode
`Distribution2 (config)# ` **`spanning-tree mode rapid-pvst`**	Enables 802.1w-based Rapid-PVST+

Core Switch (3650)

`Core> ` **`enable`**	Moves to privileged EXEC mode
`Core# ` **`configure terminal`**	Moves to global configuration mode
`Core(config)# ` **`spanning-tree mode rapid-pvst`**	Enables 802.1w-based Rapid-PVST+

CHAPTER 12

EtherChannel

This chapter provides information and commands concerning the following topics:

- EtherChannel
  - Interface modes in EtherChannel
  - Default EtherChannel configuration
  - Guidelines for configuring EtherChannel
  - Configuring Layer 2 EtherChannel
  - Configuring Layer 3 EtherChannel
  - Configuring EtherChannel load balancing
  - Configuring LACP hot-standby ports
  - Monitoring and verifying EtherChannel
- Configuration example: EtherChannel

EtherChannel

EtherChannel provides fault-tolerant, high-speed links between switches, routers, and servers. An EtherChannel consists of individual Fast Ethernet or Gigabit Ethernet links bundled into a single logical link. If a link within an EtherChannel fails, traffic previously carried over that failed link changes to the remaining links within the EtherChannel.

Interface Modes in EtherChannel

Mode	Protocol	Description
On	None	Forces the interface into an EtherChannel without Port Aggregation Protocol (PAgP) or Link Aggregation Control Protocol (LACP). Channel only exists if connected to another interface group also in On mode
Auto	PAgP (Cisco)	Places the interface into a passive negotiating state (will respond to PAgP packets but will not initiate PAgP negotiation)
Desirable	PAgP (Cisco)	Places the interface into an active negotiating state (will send PAgP packets to start negotiations)
Passive	LACP (IEEE)	Places the interface into a passive negotiating state (will respond to LACP packets but will not initiate LACP negotiation)
Active	LACP (IEEE)	Places the interface into an active negotiating state (will send LACP packets to start negotiations)

Default EtherChannel Configuration

Feature	Default Setting
Channel groups	None assigned
Port-channel logical interface	None defined
PAgP mode	No default
PAgP learn method	Aggregate-port learning on all ports
PAgP priority	128 on all ports
LACP mode	No default
LACP learn method	Aggregate-port learning on all ports
LACP port priority	32768 on all ports
LACP system priority	32768
LACP system ID	LACP system priority and the switch (or switch stack) MAC address
Load balancing	Load distribution on the switch is based on the source MAC address of the incoming packet

Guidelines for Configuring EtherChannel

- PAgP is Cisco proprietary and not compatible with LACP.
- LACP is defined in 802.3ad.
- A maximum of 48 EtherChannels are supported on a switch or switch stack.
- A single PAgP EtherChannel can be made by combining anywhere from two to eight parallel links.
- A single LACP EtherChannel can be made by combining up to 16 Ethernet ports of the same type. Up to eight ports can be active and up to eight ports can be in standby mode.
- All ports must be identical:
  - Same speed and duplex
  - Cannot mix Fast Ethernet and Gigabit Ethernet
  - Cannot mix PAgP and LACP in a single EtherChannel
  - Can have PAgP and LACP EtherChannels on the same switch, but each EtherChannel must be exclusively PAgP or LACP
  - Must all be VLAN trunk or nontrunk operational status
- All links must be either Layer 2 or Layer 3 in a single channel group.
- To create a channel in PAgP, sides must be set to one of the following:
  - Auto-Desirable
  - Desirable-Desirable

- To create a channel in LACP, sides must be set to either:
  - Active-Active
  - Active-Passive
- To create a channel without using PAgP or LACP, sides must be set to On-On.
- Do *not* configure a GigaStack gigabit interface converter (GBIC) as part of an EtherChannel.
- An interface that is already configured to be a Switched Port Analyzer (SPAN) destination port will not join an EtherChannel group until SPAN is disabled.
- Do *not* configure a secure port as part of an EtherChannel.
- Interfaces with different native VLANs cannot form an EtherChannel.
- When using trunk links, ensure that all trunks are in the same mode—Inter-Switch Link (ISL) or dot1q.
- When a group is first created, all ports follow the parameters set for the first port to be added to the group. If you change the configuration of one of the parameters, you must also make these changes to all ports in the group:
  - Allowed-VLAN list
  - Spanning-tree path cost for each VLAN
  - Spanning-tree port priority for each VLAN
  - Spanning-tree PortFast setting
- Do not configure a port that is an active or a not-yet-active member of an EtherChannel as an IEEE 802.1X port. If you try to enable IEEE 802.1X on an EtherChannel port, an error message will appear, and IEEE 802.1X is not enabled.
- For a Layer 3 EtherChannel, assign the Layer 3 address to the port-channel logical interface, not the physical ports in the channel.

Configuring Layer 2 EtherChannel

Switch(config)# **interface port-channel {number}**	Specifies the port-channel interface
Switch(config-if)# **interface {parameters}**	Once in the interface configuration mode, you can configure additional parameters
Switch(config)# **interface range fastethernet 0/1 - 4**	Moves to interface range config mode
Switch(config-if-range)# **channel-group 1 mode on**	Creates channel group 1 as an EtherChannel and assigns interfaces 01 to 04 as part of it
Switch(config-if-range)# **channel-group 1 mode desirable**	Creates channel group 1 as a PAgP channel and assigns interfaces 01 to 04 as part of it
Switch(config-if-range)# **channel-group 1 mode active**	Creates channel group 1 as an LACP channel and assigns interfaces 01 to 04 as part of it

NOTE: If you enter the **channel-group** command in the physical port interface mode without first setting a **port channel** command in global configuration mode, the port channel will automatically be created for you.

Configuring Layer 3 EtherChannel

`L3Switch(config)# interface port-channel 1`	Creates the port-channel logical interface and moves to interface config mode. Valid channel numbers are 1 to 48 for a 3560 series. For a 2960 series switch with L3 capabilities, the valid channel numbers are 1 to 6
`L3Switch(config-if)# no switchport`	Puts the interface into Layer 3 mode
`L3Switch(config-if)# ip address 172.16.10.1 255.255.255.0`	Assigns the IP address and netmask
`L3Switch(config-if)# exit`	Moves to global config mode
`L3Switch(config)# interface range fastethernet 0/20 - 24`	Moves to interface range config mode
`L3Switch(config-if)# no switchport`	Puts the interface into Layer 3 mode
`L3Switch(config-if-range)# no ip address`	Ensures that no IP addresses are assigned on the interfaces
`L3Switch(config-if-range)# channel-group 1 mode on`	Creates channel group 1 as an EtherChannel and assigns interfaces 20 to 24 as part of it
`L3Switch(config-if-range)# channel-group 1 mode desirable`	Creates channel group 1 as a PAgP channel and assigns interfaces 20 to 24 as part of it
`L3Switch(config-if-range)# channel-group 1 mode active`	Creates channel group 1 as an LACP channel and assigns interfaces 20 to 24 as part of it **NOTE:** The channel group number must match the port channel number

Configuring EtherChannel Load Balancing

`L3Switch(config)# port-channel load-balance src-mac`	Configures an EtherChannel load-balancing method. The default is **src-mac** (specifies the source MAC address of the incoming packet
Select one of the following load-distribution methods	**dst-ip**—Specifies destination host IP address **dst-mac**—Specifies destination host MAC address of the incoming packet **dst-mixed-ip-port**—Specifies destination host IP address and the TCP/UDP port **dst-port**—Specifies destination TCP/UDP port

	extended—Specifies extended load-balance methods (combination of source and destination methods beyond those available with the standard command)
	ipv6-label—Specifies the IPv6 flow label
	l3-proto—Specifies the Layer 3 protocol
	src-dst-ip—Specifies the source and destination host IP address
	src-dst-mac—Specifies the source and destination host MAC address
	src-dst-mixed-ip-port—Specifies the source and destination host IP address and TCP/UDP port
	src-ip—Specifies source host IP address
	src-mac—Specifies source host MAC address (this is the default setting)
	dst-mixed-ip-port—Specifies the source host IP address and the TCP/UDP port
	src-port—Specifies the source TCP/UDP port

Configuring LACP Hot-Standby Ports

When LACP is enabled, by default the software tries to configure the maximum number of LACP-compatible ports in a channel, up to a maximum of 16 ports. Only eight ports can be active at one time; the remaining eight links are placed into hot-standby mode. If one of the active links becomes inactive, a link in hot-standby mode becomes active in its place.

You can overwrite the default behavior by specifying the maximum number of active ports in a channel, in which case the remaining ports become hot-standby ports (if you specify only 5 active ports in a channel, the remaining 11 ports become hot-standby ports).

If you specify more than eight links for an EtherChannel group, the software automatically decides which of the hot-standby ports to make active based on LACP priority. For every link that operates in LACP, the software assigns a unique priority made up of the following (in priority order):

- LACP system priority
- System ID (the device MAC address)
- LACP port priority
- Port number

TIP: Lower numbers are better.

`Switch(config)# interface port-channel 2`	Enters interface configuration mode for port channel 2. The range for port channels is 1 to 128
`Switch(config-if)# lacp max-bundle 3`	Specifies the maximum number of LACP ports in the port-channel bundle. The range is 1 to 8
`Switch(config-if)# port-channel min-links 3`	Specifies the minimum number of member ports (in this example, 3) that must be in the link-up state and bundled in the EtherChannel for the port-channel interface to transition to the link-up state. The range for this command is 2 to 8
`Switch(config-if)# exit`	Returns to global configuration mode
`Switch(config)# lacp system-priority 32000`	Configures the LACP system priority. The range is 1 to 65535. The default is 32768. The lower the value, the higher the system priority
`Switch(config)# interface gigabitethernet 1/0/2`	Moves to interface configuration mode
`Switch(config-if)# lacp port-priority 32000`	Configures the LACP port priority. The range is 1 to 65535. The default is 32768. The lower the value, the more likely that the port will be used for LACP transmission
`Switch(config-if)# end`	Returns to privileged EXEC mode

Monitoring and Verifying EtherChannel

`Switch# show running-config`	Displays a list of what is currently running on the device
`Switch# show running-config interface fastethernet 0/12`	Displays interface fastethernet 0/12 information
`Switch# show interfaces fastethernet 0/12 etherchannel`	Displays EtherChannel information for specified interface
`Switch# show etherchannel`	Displays all EtherChannel information
`Switch# show etherchannel 1 port-channel`	Displays port channel information
`Switch# show etherchannel summary`	Displays a summary of EtherChannel information
`Switch# show interface port-channel 1`	Displays the general status of EtherChannel 1
`Switch# show lacp neighbor`	Shows LACP neighbor information
`Switch# show pagp neighbor`	Shows PAgP neighbor information
`Switch# clear pagp 1 counters`	Clears PAgP channel group 1 information
`Switch# clear lacp 1 counters`	Clears LACP channel group 1 information

Configuration Example: EtherChannel

Figure 12-1 shows the network topology for the configuration that follows, which shows how to configure EtherChannel using commands covered in this chapter.

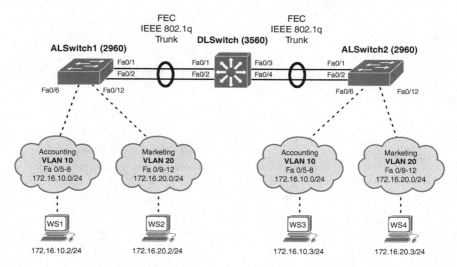

Figure 12-1 Network Topology for EtherChannel Configuration

DLSwitch (3560 or 9300)

`Switch> enable`	Moves to privileged EXEC mode
`Switch# configure terminal`	Moves to global configuration mode
`Switch(config)# hostname DLSwitch`	Sets the host name
`DLSwitch(config)# no ip domain-lookup`	Turns off DNS queries so that spelling mistakes do not slow you down
`DLSwitch(config)# vtp mode server`	Changes the switch to VTP server mode
`DLSwitch(config)# vtp domain testdomain`	Configures the VTP domain name to *testdomain*
`DLSwitch(config)# vlan 10`	Creates VLAN 10 and enters VLAN config mode
`DLSwitch(config-vlan)# name Accounting`	Assigns a name to the VLAN
`DLSwitch(config-vlan)# exit`	Returns to global config mode
`DLSwitch(config)# vlan 20`	Creates VLAN 20 and enters VLAN config mode
`DLSwitch(config-vlan)# name Marketing`	Assigns a name to the VLAN
`DLSwitch(config-vlan)# exit`	Returns to global configuration mode

DLSwitch(config)# **interface range fastethernet 0/1 - 4**	Moves to interface range config mode
DLSwitch(config-if)# **switchport trunk encapsulation dot1q**	Specifies 802.1Q tagging on the trunk link
DLSwitch(config-if)# **switchport mode trunk**	Puts the interface into permanent trunking mode and negotiates to convert the link into a trunk link
DLSwitch(config-if)# **exit**	Returns to global configuration mode
DLSwitch(config)# **interface range fastethernet 0/1 - 2**	Moves to interface range config mode
DLSwitch(config-if)# **channel-group 1 mode desirable**	Creates channel group 1 and assigns interfaces 01 to 02 as part of it
DLSwitch(config-if)# **exit**	Moves to global configuration mode
DLSwitch(config)# **interface range fastethernet 0/3 - 4**	Moves to interface range config mode
DLSwitch(config-if)# **channel-group 2 mode desirable**	Creates channel group 2 and assigns interfaces 03 to 04 as part of it
DLSwitch(config-if)# **exit**	Moves to global configuration mode
DLSwitch(config)# **port-channel load-balance dst-mac**	Configures load balancing based on destination MAC address
DLSwitch(config)# **exit**	Moves to privileged EXEC mode
DLSwitch# **copy running-config startup-config**	Saves the configuration to NVRAM

ALSwitch1 (2960 or 9200)

Switch> **enable**	Moves to privileged EXEC mode
Switch# **configure terminal**	Moves to global configuration mode
Switch(config)# **hostname ALSwitch1**	Sets host name
ALSwitch1(config)# **no ip domain-lookup**	Turns off DNS queries so that spelling mistakes do not slow you down
ALSwitch1(config)# **vtp mode client**	Changes the switch to VTP client mode
ALSwitch1(config)# **vtp domain testdomain**	Configures the VTP domain name to *testdomain*
ALSwitch1(config)# **interface range fastethernet 0/5 - 8**	Moves to interface range config mode
ALSwitch1(config-if-range)# **switchport mode access**	Sets ports 05 to 08 as access ports
ALSwitch1(config-if-range)# **switchport access vlan** 10	Assigns ports to VLAN 10
ALSwitch1(config-if-range)# **exit**	Moves to global configuration mode

ALSwitch1(config)# **interface range fastethernet 0/9 - 12**	Moves to interface range config mode
ALSwitch1(config-if-range)# **switchport mode access**	Sets ports 09 to 12 as access ports
ALSwitch1(config-if-range)# **switchport access vlan 20**	Assigns ports to VLAN 20
ALSwitch1(config-if-range)# **exit**	Moves to global configuration mode
ALSwitch1(config)# **interface range fastethernet 0/1 - 2**	Moves to interface range config mode
ALSwitch1(config-if-range)# **switchport mode trunk**	Puts the interface into permanent trunking mode and negotiates to convert the link into a trunk link
ALSwitch1(config-if-range)# **channel-group 1 mode desirable**	Creates channel group 1 and assigns interfaces 01 to 02 as part of it
ALSwitch1(config-if-range)# **exit**	Moves to global configuration mode
ALSwitch1(config)# **exit**	Moves to privileged EXEC mode
ALSwitch1# **copy running-config startup-config**	Saves the configuration to NVRAM

ALSwitch2 (2960 or 9200)

Switch> **enable**	Moves to privileged EXEC mode
Switch# **configure terminal**	Moves to global configuration mode
Switch(config)# **hostname ALSwitch2**	Sets host name
ALSwitch2(config)# **no ip domain-lookup**	Turns off DNS queries so that spelling mistakes do not slow you down
ALSwitch2(config)# **vtp mode client**	Changes the switch to VTP client mode
ALSwitch2(config)# **vtp domain testdomain**	Configures the VTP domain name to *testdomain*
ALSwitch2(config)# **interface range fastethernet 0/5 - 8**	Moves to interface range config mode
ALSwitch2(config-if-range)# **switchport mode access**	Sets ports 05 to 08 as access ports
ALSwitch2(config-if-range)# **switchport access vlan 10**	Assigns ports to VLAN 10
ALSwitch2(config-if-range)# **exit**	Moves to global configuration mode
ALSwitch2(config)# **interface range fastethernet 0/9 - 12**	Moves to interface range config mode
ALSwitch2(config-if-range)# **switchport mode access**	Sets ports 09 to 12 as access ports
ALSwitch2(config-if-range)# **switchport access vlan 20**	Assigns ports to VLAN 20

`ALSwitch2(config-if-range)# `**`exit`**	Moves to global configuration mode
`ALSwitch2(config)# `**`interface range fastethernet 0/1 - 2`**	Moves to interface range config mode
`ALSwitch2(config-if-range)# `**`switchport mode trunk`**	Puts the interface into permanent trunking mode and negotiates to convert the link into a trunk link
`ALSwitch2(config-if-range)# `**`channel-group 2 mode desirable`**	Creates channel group 2 and assigns interfaces 01 to 02 as part of it **NOTE:** Although the local channel group number does not have to match the channel group number on a neighboring switch, the numbers are often chosen to be the same for ease of management and documentation purposes.
`ALSwitch2(config-if-range)# `**`exit`**	Moves to global configuration mode
`ALSwitch2(config)# `**`exit`**	Moves to privileged EXEC mode
`ALSwitch2# `**`copy running-config startup-config`**	Saves the configuration to NVRAM

CHAPTER 13

Cisco Discovery Protocol (CDP) and Link Layer Discovery Protocol (LLDP)

This chapter provides information and commands concerning the following topic:

- Cisco Discovery Protocol
- Configuring CDP
- Verifying and Troubleshooting CDP
- CDP Design Tips
- Link Layer Discovery Protocol (802.1AB)
- Configuring LLDP (802.1AB)
- Verifying and troubleshooting LLDP

Cisco Discovery Protocol

Cisco Discovery Protocol (CDP) is a Cisco proprietary Layer 2 protocol. It is media and protocol independent and runs on all Cisco-manufactured equipment including routers, bridges, access servers, and switches. CDP is primarily used to obtain protocol addresses of neighboring devices and discover the platform of those devices. CDP can also be used to display information about the interfaces that your device uses.

Configuring CDP

NOTE: CDP is enabled on all Cisco devices by default.

Router(config)# **cdp holdtime *x***	Changes the length of time to keep CDP packets
Router(config)# **cdp timer *x***	Changes how often CDP updates are sent
Router(config)# **cdp run**	Enables CDP globally (on by default)
Router(config)# **no cdp run**	Turns off CDP globally
Router(config-if)# **cdp enable**	Enables CDP on a specific interface
Router(config-if)# **no cdp enable**	Turns off CDP on a specific interface

Verifying and Troubleshooting CDP

Router# **show cdp**	Displays global CDP information (such as timers)
Router# **show cdp neighbors**	Displays information about neighbors
Router# **show cdp neighbors detail**	Displays more detail about neighbor devices
Router# **show cdp entry word**	Displays information about the device named *word*
Router# **show cdp entry ***	Displays information about all devices
Router# **show cdp interface**	Displays information about interfaces that have CDP running
Router# **show cdp interface x**	Displays information about specific interface *x* running CDP
Router# **show cdp traffic**	Displays traffic information—packets in/out/version
Router# **clear cdp counters**	Resets traffic counters to 0
Router# **clear cdp table**	Deletes the CDP table
Router# **debug cdp adjacency**	Monitors CDP neighbor information
Router# **debug cdp events**	Monitors all CDP events
Router# **debug cdp ip**	Monitors CDP events specifically for IP
Router# **debug cdp packets**	Monitors CDP packet-related information

CDP Design Tips

CAUTION: Although CDP is necessary for some management applications, CDP should still be disabled in some instances.

Disable CDP globally if

- CDP is not required at all.
- The device is located in an insecure environment.

Use the command **no cdp run** to disable CDP globally:

```
RouterOrSwitch(config)# no cdp run
```

Disable CDP on any interface if

- Management is not being performed.
- The switch interface is a nontrunk interface.
- The interface is connected to a nontrusted network.

Use the interface configuration command **no cdp enable** to disable CDP on a specific interface:

```
RouterOrSwitch(config)# interface fastethernet 0/1
RouterOrSwitch(config-if)# no cdp enable
```

Link Layer Discovery Protocol (802.1AB)

Link Layer Discovery Protocol (LLDP) is an industry-standard alternative to CDP. LLDP is a neighbor discovery protocol that is used for network devices to advertise information about themselves to other devices on the network. This protocol runs over the data link layer, which allows two systems running different network layer protocols to learn about each other.

NOTE: LLDP works on Ethernet type interfaces only.

NOTE: A switch stack appears as a single switch in the network. Therefore, LLDP discovers the switch stack, not the individual stack members.

Configuring LLDP (802.1AB)

Switch(config)# **lldp run**	Enables LLDP globally on the switch
Switch(config)# **no lldp run**	Disables LLDP globally on the switch
Switch(config)# **lldp holdtime 180**	Specifies the amount of time a receiving device should hold the information sent by another device before discarding it. The default value is 120 seconds. The range is 0 to 65,535 seconds
Switch(config)# **lldp timer 60**	Sets the transmission frequency of LLDP updates in seconds. The default value is 30 seconds. The range is 5 to 65,534 seconds
Switch(config)# **lldp reinit 3**	(Optional) Specifies the delay time in seconds for LLDP to initialize on any interface. The range is 2 to 5 seconds. The default is 2 seconds
Switch(config)# **interface fastethernet0/1**	Specifies the interface on which you are enabling or disabling LLDP and enters interface configuration mode
Switch(config-if)# **lldp transmit**	Enables the interface to send LLDP
Switch(config-if)# **lldp receive**	Enables the interface to receive LLDP
Switch(config-if)# **no lldp transmit**	No LLDP packets are sent on the interface
Switch(config-if)# **no lldp receive**	No LLDP packets are received on the interface

Verifying and Troubleshooting LLDP

Switch# **clear lldp counters**	Resets the traffic counters to 0
Switch# **clear lldp table**	Deletes the LLDP table of information about neighbors
Switch# **debug lldp packets**	Enables debugging of LLDP packets. Use the **no** form of this command to disable debugging
Switch# **show lldp**	Displays global information, such as frequency of transmissions, the holdtime for packets being sent, and the delay time for LLDP to initialize on an interface
Switch# **show lldp entry** *entry-name*	Displays information about a specific neighbor **TIP:** You can enter an asterisk (*) to display all neighbors, or you can enter the name of the neighbor about which you want information
Switch# **show lldp interface** *[interface-id]*	Displays information about interfaces where LLDP is enabled. You can limit the display to the interface about which you want information
Switch# **show lldp neighbors** *[interface-id]* *[detail]*	Displays information about neighbors, including device type, interface type and number, holdtime settings, capabilities, and port ID
Switch# **show lldp traffic**	Displays LLDP counters, including the number of packets sent and received, number of packets discarded, and number of unrecognized Type Length Value (TLV) fields

CHAPTER 14

Configuring a Cisco Router

This chapter provides information and commands concerning the following topics:

- Router modes
- Entering global configuration mode
- Configuring a router, specifically
    - Device name
    - Passwords
    - Password encryption
    - Interface name
    - Moving between interfaces
    - Configuring a serial interface
    - Assigning an IPv4 address to a Fast Ethernet interface
    - Assigning an IPv4 address to a Gigabit Ethernet interface
    - Assigning IPv6 addresses to interfaces
    - Creating a message-of-the-day (MOTD) banner
    - Creating a login banner
    - Mapping a local host name to a remote IP address
    - The **no ip domain-lookup** command
    - Working with DNS on a router
    - The **logging synchronous** command
    - The **exec-timeout** command
    - Saving configurations
    - Erasing configurations
    - The **write** command
- Verifying your configuration using **show** commands
- EXEC commands in configuration mode: the **do** command
- Configuration example: basic router configuration

Router Modes

`Router>`	User mode
`Router#`	Privileged EXEC mode (also known as EXEC-level mode)
`Router(config)#`	Global configuration mode
`Router(config-if)#`	Interface mode
`Router(config-subif)#`	Subinterface mode
`Router(config-line)#`	Line mode
`Router(config-router)#`	Router configuration mode

TIP: There are other modes than these. Not all commands work in all modes. Be careful. If you type in a command that you know is correct—**show running-config**, for example—and you get an error, make sure that you are in the correct mode.

Entering Global Configuration Mode

`Router>`	Limited viewing of configuration. You cannot make changes in this mode
`Router>` **`enable`**	Moves to privileged EXEC mode
`Router#`	You can see the configuration and move to make changes
`Router#` **`configure terminal`** `Router(config)#`	Moves to global configuration mode. This prompt indicates that you can start making changes

Configuring a Router Name

This command works on both routers and switches

`Router(config)#` **`hostname Cisco`**	The name can be any word you choose. The name should start with a letter and contain no spaces
`Cisco(config)#`	Notice that the name of the router has changed from the default Router to Cisco

Configuring Passwords

These commands work on both routers and switches.

`Router(config)#` **`enable password cisco`**	Sets **enable** password to *cisco*
`Router(config)#` **`enable secret class`**	Sets **enable secret** password to *class*
`Router(config)#` **`line console 0`**	Enters console line mode
`Router(config-line)#` **`password console`**	Sets console line mode password to *console*

`Router(config-line)# login`	Enables password checking at login
`Router(config)# line vty 0 4`	Enters vty line mode for all five vty lines
`Router(config-line)# password telnet`	Sets vty password to *telnet*
`Router(config-line)# login`	Enables password checking at login
`Router(config)# line aux 0`	Enters auxiliary line mode **NOTE:** This is not available on Cisco switches
`Router(config-line)# password backdoor`	Sets auxiliary line mode password to *backdoor*
`Router(config-line)# login`	Enables password checking at login

CAUTION: The **enable secret** *password* is encrypted by default. The **enable** *password* is not. For this reason, recommended practice is that you *never* use the **enable** *password* command. Use only the **enable secret** *password* command in a router or a switch configuration. You cannot set both **enable secret** *password* and **enable** *password* to the same password. Doing so defeats the use of encryption.

Password Encryption

`Router(config)# service password-encryption`	Clear text passwords will be hidden using a weak encryption algorithm
`Router(config)# enable password cisco`	Sets enable password to *cisco*
`Router(config)# line console 0`	Moves to console line mode
`Router(config-line)# password Cisco`	Continue setting passwords as above
	. . .
`Router(config)# no service password-encryption`	Turns off password encryption

CAUTION: If you have turned on **service password-encryption**, used it, and then turned it off, any passwords that you have encrypted stay encrypted. New passwords remain unencrypted.

Interface Names

One of the biggest problems that new administrators face is the interface names on the different models of routers. With all the different Cisco devices in production networks today, some administrators are becoming confused about the names of their interfaces. Using Cisco devices that are no longer in production but are still valuable in a lab or classroom setting can also complicate matters. Older devices are still a great (and inexpensive) way to learn the basics (and in some cases the more advanced methods) of router configuration.

The following chart is a sample of *some* of the different interface names for various routers. This is by no means a complete list. Refer to the hardware guide of the specific router that you are working on to see the various combinations, or use the following command to see which interfaces are installed on your particular router:

`router# show ip interface brief`

NOTE: An "on-board" port is a fixed port that is built directly into the motherboard. A "slot" is used to expand port density of a device by inserting a module that plugs into the motherboard. A module may contain several ports. Depending on the router, you may have no slots or many.

Router Model	Port Location/ Slot Number	Slot/Port Type	Slot Numbering Range	Example
2501	On board	Ethernet	Interface-type number	ethernet0 (e0)
	On board	Serial	Interface-type number	serial0 (s0) and s1
2514	On board	Ethernet	Interface-type number	e0 and e1
	On board	Serial	Interface-type number	s0 and s1
1721	On board	Fast Ethernet	Interface-type number	fastethernet0 (fa0)
	Slot 0	Wireless Access Controller (WAC)	Interface-type number	s0 and s1
1760	On board	Fast Ethernet	Interface-type 0/port	fa0/0
	Slot 0	WAN Interface Card (WIC)/ Voice Interface Card (VIC)	Interface-type 0/port	s0/0 and s0/1 v0/0 and v0/1
	Slot 1	WIC/VIC	Interface-type 1/port	s1/0 and s1/1 v1/0 and v1/1
	Slot 2	VIC	Interface-type 2/port	v2/0 and v2/1
	Slot 3	VIC	Interface-type 3/port	v3/0 and v3/1
2610	On board	Ethernet	Interface-type 0/port	e0/0
	Slot 0	WIC (serial)	Interface-type 0/port	s0/0 and s0/1
2611	On board	Ethernet	Interface-type 0/port	e0/0 and e0/1
	Slot 0	WIC (serial)	Interface-type 0/port	s0/0 and s0/1
2620	On board	Fast Ethernet	Interface-type 0/port	fa0/0
	Slot 0	WIC (serial)	Interface-type 0/port	s0/0 and s0/1
2621	On board	Fast Ethernet	Interface-type 0/port	fa0/0 and fa0/1
	Slot 0	WIC (serial)	Interface-type 0/port	s0/0 and s0/1
1841	On board	Fast Ethernet	Interface-type 0/port	fa0/0 and fa0/1

Router Model	Port Location/ Slot Number	Slot/Port Type	Slot Numbering Range	Example
	Slot 0	High-speed WAN Interface Card (HWIC)/WIC/ Voice WAN Interface Card (VWIC)	Interface-type 0/slot/ port	s0/0/0 and s0/0/1
	Slot 1	HWIC/WIC/ VWIC	Interface-type 0/slot/ port	s0/1/0 and s0/1/1
2801	On board	Fast Ethernet	Interface-type 0/port	fa0/0 and fa0/1
	Slot 0	VIC/VWIC (voice only)	Interface-type 0/slot/ port	voice0/0/0– voice0/0/3
	Slot 1	HWIC/WIC/ VWIC	Interface-type 0/slot/ port	0/1/0–0/1/3 (single-wide HWIC) 0/1/0– 0/1/7 (double-wide HWIC)
	Slot 2	WIC/VIC/VWIC	Interface-type 0/slot/ port	0/2/0–0/2/3
	Slot 3	HWIC/WIC/ VWIC	Interface-type 0/slot/ port	0/3/0–0/3/3 (single-wide HWIC) 0/3/0– 0/3/7 (double-wide HWIC)
2811	Built in to chassis front	USB	Interface-type port	usb0 and usb1
	Built in to chassis rear	Fast Ethernet Gigabit Ethernet	Interface-type 0/port	fa0/0 and fa0/1 gi0/0 and gi0/1
	Slot 0	HWIC/HWIC-D/ WIC/VWIC/VIC	Interface-type 0/slot/ port	s0/0/0 and s0/0/1 fa0/0/0 and 0/0/1
	Slot 1	HWIC/High-Speed WAN Interface Card-Double-wide (HWIC-D)/ WIC/VWIC/VIC	Interface-type 0/slot/ port	s0/1/0 and s0/1/1 fa0/1/0 and 0/1/1

Router Model	Port Location/ Slot Number	Slot/Port Type	Slot Numbering Range	Example
	NME slot	Network Module (NM)/Network Module Enhanced (NME)	Interface-type 1/port	gi1/0 and gi1/1 s1/0 and s1/1
1941 / 1941w	On board	Gigabit Ethernet	Interface-type 0/port	gi0/0 and gi0/1
	Slot 0	Enhanced High-Speed WAN Interface Card (EHWIC)	Interface-type 0/slot/ port	s0/0/0 and s0/0/1
	Slot 1	EHWIC	Interface-type 0/slot/ port	s0/1/0 and s0/1/1
	Built in to chassis back	USB	Interface-type port	usb0 and usb 1
2901 2911	On board	Gigabit Ethernet	Interface-type 0/port	gi0/0 and gi0/1 gi0/2 (2911 only)
	Slot 0	EHWIC	Interface-type 0/slot/ port	s0/0/0 and s0/0/1
	Slot 1	EHWIC	Interface-type 0/slot/ port	s0/1/0 and s0/1/1
	Slot 2	EHWIC	Interface-type 0/slot/ port	s0/2/0 and s0/2/1
	Slot 3	EHWIC	Interface-type 0/slot/ port	s0/3/0 and s0/3/1
	Built in to chassis back	USB	Interface-type port	usb0 and usb 1
4221 / 4321	On board	Gigabit Ethernet	Interface-type 0/slot/ port	gi0/0/0 and gi0/0/1
		Gigabit Ethernet	Interface-type 0/slot/ port (SFP fiber-optic port)	gi0/0/0
		NOTE: Only one of the RJ45 Gi0/0/0 or SFP Gi0/0/0 can be used, as they share the same interface name (Gi0/0/0)		

Router Model	Port Location/ Slot Number	Slot/Port Type	Slot Numbering Range	Example
	Slot 1	NIMs (Network Interface Modules) Both serial and Ethernet cards are available for NIM slots	Interface-type 0/slot/ port	s0/1/0 and s0/1/1 or gi0/1/0 and gi0/1/1
	Slot 2	NIMs Both serial and Ethernet cards are available for NIM slots	Interface-type 0/slot/port	s0/2/0 and s0/2/1 or gi0/2/0 and gi0/2/1

Moving Between Interfaces

When moving between interfaces, you have two options. The first option, shown on the left side of the following table, exits out of interface mode back to global configuration mode, and then enters into a new interface mode. In this scenario, the prompt changes and you see the movement. The second option, shown on the right side of the table, moves directly from one interface mode to the second interface mode. In this case, the prompt does not change, even though you are in a new interface mode.

CAUTION: You do not want to put the configuration for one interface on a different interface.

Exiting One Interface and Entering a New Interface		Moving Directly Between Interfaces	

`Router(config)#`
`interface`
`serial 0/0/0`
	Moves to serial interface configuration mode	`Router(config)#`
		`interface serial`
		`0/0/0`
			Moves to serial interface configuration mode

`Router`
`(config-if)#`
`exit`
	Returns to global configuration mode	`Router`
		`(config-if)#`
		`interface`
		`fastethernet 0/0`
			Moves directly to Fast Ethernet 0/0 configuration mode

`Router(config)#`
`interface`
`fastethernet 0/0`
	Moves to Fast Ethernet interface configuration mode	`Router`
		`(config-if)#`
			In Fast Ethernet 0/0 configuration mode now

`Router` `(config-if)#`	In Fast Ethernet 0/0 configuration mode now	`Router` `(config-if)#`	Prompt does not change; be careful

Configuring a Serial Interface

Router(config)# **interface serial 0/0/0**	Moves to serial interface 0/0/0 configuration mode
Router(config-if)# **description Link to ISP**	Optional descriptor of the link is locally significant
Router(config-if)# **ip address 192.168.10.1 255.255.255.0**	Assigns address and subnet mask to interface
Router(config-if)# **clock rate 2000000**	Assigns a clock rate for the interface
Router(config-if)# **no shutdown**	Turns interface on

TIP: The **clock rate** command is used only on a *serial* interface that has a *DCE* cable plugged into it. There must be a clock rate on every serial link between routers. It does not matter which router has the DCE cable plugged into it or which interface the cable is plugged into. Serial 0/0/0 on one router can be plugged into Serial 0/0/1 on another router.

NOTE: Serial connections are rapidly being removed from networks because Ethernet connections are faster and not reliant on clocking rates. In this book, serial interfaces are used to distinguish between WAN connections and LAN connections (which are shown using Ethernet interfaces).

Assigning an IPv4 Address to a Fast Ethernet Interface

Router(config)# **interface fastethernet 0/0**	Moves to Fast Ethernet 0/0 interface configuration mode
Router(config-if)# **description Accounting LAN**	Optional descriptor of the link is locally significant
Router(config-if)# **ip address 192.168.20.1 255.255.255.0**	Assigns address and subnet mask to interface
Router(config-if)# **no shutdown**	Turns interface on

Assigning an IPv4 Address to a Gigabit Ethernet Interface

Router(config)# **interface gigabitethernet 0/0/0**	Moves to gigabitethernet 0/0/0 interface configuration mode
Router(config-if)# **description Human Resources LAN**	Optional descriptor of the link is locally significant
Router(config-if)# **ip address 192.168.30.1 255.255.255.0**	Assigns an address and subnet mask to interface
Router(config-if)# **no shutdown**	Turns interface on

Assigning IPv6 Addresses to Interfaces

`Router(config)# ipv6` `unicast-routing`	Enables the forwarding of IPv6 unicast datagrams globally on the router
`Router(config)# interface` `gigabitethernet 0/0/0`	Moves to interface configuration mode
`Router(config-if)#` `ipv6 enable`	Automatically configures an IPv6 link-local address on the interface and enables IPv6 processing on the interface **NOTE:** The link-local address that the **ipv6 enable** command configures can be used only to communicate with nodes on the same broadcast segment
`Router(config-if)# ipv6` `address autoconfig`	Router configures itself with a link-local address using stateless autoconfiguration
`Router(config-if)# ipv6` `address 2001::1/64`	Configures a global IPv6 address on the interface and enables IPv6 processing on the interface

`Router(config-if)# ipv6`
`address 2001:db8:0:1::/64`
`eui-64`
	Configures a global IPv6 address with an interface identifier in the low-order 64 bits of the IPv6 address

`Router(config-if)# ipv6`
`address fe80::260:3eff:`
`fe47:1530/ 64 link-local`
	Configures a specific link-local IPv6 address on the interface instead of the one that is automatically configured when IPv6 is enabled on the interface

`Router(config-if)# ipv6` `unnumbered` type/number	Specifies an unnumbered interface and enables IPv6 processing on the interface. The global IPv6 address of the interface specified by *type/number* will be used as the source address

Creating a Message-of-the-Day Banner

`Router(config)# banner motd ^`
`Building Power will be`
`interrupted next Tuesday`
`evening from 8 - 10 PM. ^`
`Router(config)#`
	^ is being used as a *delimiting character*. The delimiting character must surround the banner message and can be any character as long as it is not a character used within the body of the message

TIP: The message-of-the-day (MOTD) banner is displayed on all terminals and is useful for sending messages that affect all users. Use the **no banner motd** command to disable the MOTD banner. The MOTD banner displays before the login prompt and the login banner, if one has been created, if you are connected via the console or through Telnet. If you are connecting using SSH, the MOTD banner appears after the SSH connection.

Creating a Login Banner

`Router(config)# banner login`
`^Authorized Personnel Only!`
`Please enter your username`
`and password. ^`
`Router(config)#`
	^ is being used as a *delimiting character*. The delimiting character must surround the banner message and can be any character as long as it is not a character used within the body of the message

TIP: The login banner displays before the username and password login prompts. Use the **no banner login** command to disable the login banner. The MOTD banner displays before the login banner.

Mapping a Local Host Name to a Remote IP Address

`Router(config)#`
`ip host london`
`172.16.1.3`
	Assigns a locally significant host name to the IP address. After this assignment, you can use the host name rather than an IP address when trying to telnet or ping to that address

`Router# ping london`
`=`
`Router# ping`
`172.16.1.3`
	Both commands execute the same objective: sending a ping to address 172.16.1.3

TIP: When in user EXEC or privileged EXEC mode, commands that do not match a valid command default to Telnet. Therefore, you can use a host name mapping to Telnet to a remote device:

   `Router# london` = `Router# telnet london` = `Router# telnet 172.16.1.3`

The no ip domain-lookup Command

`Router(config)# no ip domain-lookup` `Router(config)#`	Turns off trying to automatically resolve an unrecognized command to a local host name

TIP: Ever type in a command incorrectly and end up having to wait for what seems to be a minute or two as the router tries to translate your command to a domain server of 255.255.255.255? When in user EXEC or privileged EXEC modes, commands that do not match a valid command default to Telnet. Also, the router is set by default to try to resolve any word that is not a command to a Domain Name System (DNS) server at address 255.255.255.255. If you are not going to set up DNS, turn off this feature to save you time as you type, especially if you are a poor typist.

NOTE: In some newer versions of the IOS, this command might not have a hyphen in it: the command is **no ip domain lookup**.

Working with DNS on a Router

The reason I created the *CCNA Portable Command Guide* is because I am a poor typist and I was always waiting for my spelling mistakes to be resolved through a DNS lookup.

If you do not have a DNS server configured, all of those spelling mistakes take time to be resolved. This is why I was so happy to discover the **no ip domain-lookup** command!

But what happens if you have a DNS server configured (using the **ip name-server** command) and **no ip domain-lookup** configured? Your DNS server is now useless because it will not be used.

A more proper way of doing things would be to configure your DNS server using the **ip name-server** command, and then go to all of your lines (con 0, aux 0, vty 0 15), and deactivate the automatic action of telnetting into all "words" that look like host names. The Cisco IOS Software accepts a host name entry at the EXEC prompt as a Telnet command. If you enter the host name incorrectly, the Cisco IOS Software interprets the entry as an incorrect Telnet command and provides an error message indicating that the host does not exist. The **transport preferred none** command disables this option so that if you enter a command incorrectly at the EXEC prompt, the Cisco IOS Software does not attempt to make a Telnet connection.

Router(config)# **line console 0**	Moves to line console configuration mode
Router(config-line)# **transport preferred none**	Deactivates automatic action of telnetting into words that look like host names (your spelling mistakes that do not look like commands)
Router(config-line)# **line aux 0**	Moves to line auxiliary configuration mode
Router(config-line)# **transport preferred none**	Deactivates automatic action of telnetting into words that look like host names (your spelling mistakes that do not look like commands)
Router(config-line)# **line vty 0 15**	Moves to virtual Telnet lines 0 through 15
Router(config-line)# **transport preferred none**	Deactivates automatic action of telnetting into words that look like host names (your spelling mistakes that do not look like commands)

Now if you make a spelling mistake at the command prompt, you will be given an error, as opposed to waiting for your mistake to be resolved through a DNS lookup.

Router# **confog** ^	Spelling mistake entered
% Invalid input detected at ^ marker Router#	No DNS lookup. Returned to prompt

The logging synchronous Command

Router(config)# **line console 0**	Moves to line console configuration mode
Router(config-line)# **logging synchronous**	Turns on synchronous logging. Information items sent to the console do not interrupt the command you are typing. The command is moved to a new line

TIP: Ever try to type in a command and an informational line appears in the middle of what you were typing? Lose your place? Do not know where you are in the command, so you just press Enter and start all over? The **logging synchronous** command tells the router that if any informational items get displayed on the screen, your prompt and command line should be moved to a new line, so as not to confuse you. The informational line does not get inserted into the middle of the command you are trying to type. If you were to continue typing, the command would execute properly, even though it looks wrong on the screen.

TIP: If you do not set the **logging synchronous** command and you are in a situation where your command being entered is interrupted by informational items being displayed on the screen, you can use the keyboard shortcut of Ctrl-R to bring your command to the next line without the message interfering with the command.

The exec-timeout Command

`Router(config)# `**`line console 0`**	Moves to line console configuration mode
`Router(config-line)# exec-timeout 0 0`	Sets the limit of idle connection time, after which the console automatically logs off. A setting of **0 0** (minutes seconds) means the console never logs off—you have disabled the timeout
	Using the command without the seconds parameter will also work to disable the timeout:
	Router(config-line)#**exec-timeout 0**
`Router(config-line)#`	

TIP: The command **exec-timeout 0** is great for a lab environment because the console never logs out, regardless of how long the connection remains idle. This is considered to be bad security and is dangerous in the real world. The default for the **exec-timeout** command is 10 minutes and zero (0) seconds (**exec-timeout 10 0**) of idle connection time.

Saving Configurations

`Router# `**`copy running-config startup-config`**	Saves the running configuration to local NVRAM. You will be prompted for a destination filename
`Router# `**`copy running-config tftp`**	Saves the running configuration remotely to a TFTP server. You will be prompted to enter in the IP address of the TFTP server

Erasing Configurations

`Router# `**`erase startup-config`**	Deletes the startup configuration file from NVRAM. You will be prompted to confirm this action as a safety precaution

TIP: The running configuration is still in dynamic memory. Reload the router to clear the running configuration.

The write Command

Router# **write**	Saves the running configuration to local NVRAM. You are not prompted for a destination file name
Router# **write memory**	Saves the running configuration to local NVRAM. You are not prompted for a destination file name
Router# **write erase**	Deletes the startup configuration file from NVRAM. You will be prompted to confirm this action as a safety precaution
Router# **write network**	Saves the running configuration remotely to a TFTP server. You will be given a message showing this command has been replaced with the **copy running-config <url>** command

NOTE: The **write** command existed before the **copy running-config startup-config** and **erase startup-config** commands. Although the **write** command was officially deprecated some time ago, it still works in many versions of the Cisco IOS Software. However, it does not work on all devices and platforms—for example, it does not work with the Nexus platform.

Verifying Your Configurations Using show Commands

Router# **show ?**	Lists all **show** commands available
Router# **show arp**	Displays the Address Resolution Protocol (ARP) table
Router# **show clock**	Displays time set on device
Router# **show controllers serial 0/0/0**	Displays statistics for interface hardware. Statistics display if the clock rate is set and if the cable is Data Communications Equipment (DCE), data terminal equipment (DTE), or not attached
Router# **show flash**	Displays info about flash memory
Router# **show history**	Displays the history of commands used at privileged EXEC level
Router# **show hosts**	Displays the local host-to-IP address cache. These are the names and addresses of hosts on the network to which you can connect
Router# **show interface serial 0/0/0**	Displays statistics for a specific interface (in this case, serial 0/0/0)
Router# **show interfaces**	Displays statistics for all interfaces
Router# **show ip interface brief**	Displays a summary of all interfaces, including status and IP address assigned
Router# **show ip protocols**	Displays the parameters and the current state of the active IPv4 routing protocol processes
Router# **show ipv6 interface brief**	Displays a summary of all interfaces, including status and IPv6 address assigned
Router# **show ipv6 protocols**	Displays the parameters and the current state of the active IPv6 routing protocol processes

Router# `show protocols`	Displays the status of configured Layer 3 protocols
Router# `show running-config`	Displays the configuration currently running in RAM
Router# `show startup-config`	Displays the configuration saved in NVRAM
Router# `show users`	Displays all users connected to the device
Router# `show version`	Displays info about loaded software version

EXEC Commands in Configuration Mode: The do Command

Router(config)# `do show running-config`	Executes the privileged-level **show running-config** command while in global configuration mode
Router(config)#	The router remains in global configuration mode after the command has been executed

TIP: The **do** command is useful when you want to execute EXEC commands, such as **show**, **clear**, or **debug**, while remaining in global configuration mode or in any configuration submode. You cannot use the **do** command to execute the **configure terminal** command because it is the **configure terminal** command that changes the mode to global configuration mode.

Configuration Example: Basic Router Configuration

Figure 14-1 illustrates the network topology for the configuration that follows, which shows a basic router configuration using the commands covered in this chapter.

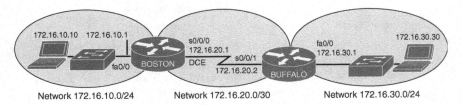

Figure 14-1 Network Topology for Basic Router Configuration

Boston Router

Router> `enable`	Enters privileged EXEC mode
Router# `configure terminal`	Enters global configuration mode
Router(config)# `hostname Boston`	Sets the router name to *Boston*
Boston(config)# `no ip domain-lookup`	Turns off name resolution on unrecognized commands (spelling mistakes)

`Boston(config)# banner login`
`#This is the Boston Router.`
`Authorized Access Only`
`#`
	Creates a login banner

`Boston(config)# enable secret` `cisco`	Enables secret password set to *cisco*
`Boston(config)# service` `password-encryption`	Clear text passwords will be hidden using a weak encryption algorithm
`Boston(config)# line console 0`	Enters line console mode
`Boston(config-line)# logging` `synchronous`	Commands will not be interrupted by unsolicited messages
`Boston(config-line)# password` `class`	Sets the password to *class*
`Boston(config-line)# login`	Enables password checking at login
`Boston(config-line)# line vty` `0 4`	Moves to virtual Telnet lines 0 through 4
`Boston(config-line)# password` `class`	Sets the password to *class*
`Boston(config-line)# login`	Enables password checking at login
`Boston(config-line)# line aux 0`	Moves to line auxiliary mode
`Boston(config-line)# password` `class`	Sets the password to *class*
`Boston(config-line)# login`	Enables password checking at login
`Boston(config-line)# exit`	Moves back to global configuration mode
`Boston(config)# no service` `password-encryption`	Turns off password encryption
`Boston(config)# interface` `fastethernet 0/0`	Moves to interface Fast Ethernet 0/0 configuration mode
`Boston(config-if)# description` `Engineering LAN`	Sets locally significant description of the interface
`Boston(config-if)# ip address` `172.16.10.1 255.255.255.0`	Assigns an IP address and subnet mask to the interface
`Boston(config-if)# no shutdown`	Turns on the interface
`Boston(config-if)# interface` `serial 0/0/0`	Moves directly to interface serial 0/0/0 configuration mode
`Boston(config-if)# description` `Link to Buffalo Router`	Sets a locally significant description of the interface
`Boston(config-if)# ip address` `172.16.20.1 255.255.255.252`	Assigns an IP address and subnet mask to the interface
`Boston(config-if)# clock rate` `56000`	Sets a clock rate for serial transmission. The DCE cable must be plugged into this interface

Boston(config-if)# **no shutdown**	Turns on the interface
Boston(config-if)# **exit**	Moves back to global configuration mode
Boston(config)# **ip host buffalo 172.16.20.2**	Sets a local host name resolution to remote IP address 172.16.20.2
Boston(config)# **exit**	Moves back to privileged EXEC mode
Boston# **copy running-config startup-config**	Saves the running configuration to NVRAM

Static Routing

This chapter provides information and commands concerning the following topics:

- Configuring an IPv4 static route
- Static routes and recursive lookups in IPv4
- The **permanent** keyword
- Floating static routes in IPv4 and administrative distance
- Configuring an IPv4 default route
- Verifying IPv4 static routes
- Configuration example: IPv4 static routes
- Configuring an IPv6 static route
- Floating static routes in IPv6
- Configuring an IPv6 default route
- Verifying IPv6 static routes

Configuring an IPv4 Static Route

When using the **ip route** command, you can identify where packets should be routed in two ways:

- The next-hop address
- The exit interface

Both ways are shown in the "Configuration Example: IPv4 Static Routes" and the "Configuring an IPv4 Default Route" sections.

`Router(config)# ip`
`route 172.16.20.0`
`255.255.255.0`
`172.16.10.2`
	172.16.20.0 = destination network
	255.255.255.0 = subnet mask
	172.16.10.2 = next-hop address
	Read this to say, "To get to the destination network of 172.16.20.0, with a subnet mask of 255.255.255.0, send all packets to 172.16.10.2"

`Router(config)# ip`
`route 172.16.20.0`
`255.255.255.0`
`serial 0/0/0`
	172.16.20.0 = destination network
	255.255.255.0 = subnet mask
	Serial 0/0/0 = exit interface
	Read this to say, "To get to the destination network of 172.16.20.0, with a subnet mask of 255.255.255.0, send all packets out interface serial 0/0/0"

Static Routes and Recursive Lookups

A static route that uses a next-hop address (intermediate address) causes the router to look at the routing table twice: once when a packet first enters the router and the router looks up the entry in the table, and a second time when the router has to resolve the location of the intermediate address.

For point-to-point links, always use an exit interface in your static route statements:

`Router(config)# ip route 192.168.10.0 255.255.255.0 serial 0/0/0`

For broadcast links such as Ethernet, Fast Ethernet, or Gigabit Ethernet, use both an exit interface and an intermediate address:

`Router(config)# ip route 192.168.10.0 255.255.255.0 fastethernet 0/0`
`192.138.20.2`

This saves the router from having to do a recursive lookup for the intermediate address of 192.168.20.2, knowing that the exit interface is FastEthernet 0/0.

Try to avoid using static routes that reference only intermediate addresses.

The permanent Keyword

Without the **permanent** keyword in a static route statement, a static route is removed if an interface goes down. A downed interface causes the directly connected network and any associated static routes to be removed from the routing table. If the interface comes back up, the routes are returned.

Adding the **permanent** keyword to a static route statement keeps the static routes in the routing table even if the interface goes down and the directly connected networks are removed. You cannot get to these routes—the interface is down—but the routes remain in the table. The advantage to this is that when the interface comes back up, the static routes do not need to be reprocessed and placed back into the routing table, thus saving time and processing power.

When a static route is added or deleted, this route, along with all other static routes, is processed in 1 second. Before Cisco IOS Software Release 12.0, this processing time was 5 seconds.

The routing table processes static routes every minute to install or remove static routes according to the changing routing table.

To specify that the route will not be removed, even if the interface shuts down, enter the following command, for example:

```
Router(config)# ip route 172.16.20.0 255.255.255.0 172.16.10.2
  permanent
```

Floating Static Routes in IPv4 and Administrative Distance

To specify that an administrative distance (AD) of 200 has been assigned to a given route, enter the following command, for example:

```
Router(config)# ip route 172.16.20.0 255.255.255.0 172.16.10.2 200
```

By default, a static route is assigned an AD of 1. AD rates the "trustworthiness" of a route. AD is a number from 0 to 255 (or 254 for IPv6), where 0 is absolutely trusted and 255 (254 for IPv6) cannot be trusted at all. Therefore, an AD of 1 is an extremely reliable rating, with only an AD of 0 being better. An AD of 0 is assigned to a directly connected route. Table 15-1 lists the AD for each type of route.

TABLE 15-1 Administrative Distances

Route Type	AD IPv4	AD IPv6
Connected	0	0
Static	1	1
Enhanced Interior Gateway Routing Protocol (EIGRP) summary route	5	5
External Border Gateway Protocol (eBGP)	20	20
EIGRP (internal)	90	90
Open Shortest Path First Protocol (OSPF)	110	110
Intermediate System-to-Intermediate System Protocol (IS-IS)	115	115
Routing Information Protocol (RIP)	120	120
External Gateway Protocol (EGP) (no longer supported by Cisco IOS)	140	140
On-Demand Routing	160	160
EIGRP (external)	170	170
Internal Border Gateway Protocol (iBGP) (external)	200	200
Unknown or unbelievable	**255** (Does not pass traffic)	**254** (Does not pass traffic)

By default, a static route is always used rather than a routing protocol. By adding an AD number to your statement, however, you can effectively create a backup route to your routing protocol. If your network is using EIGRP and you need a backup route, add a static route with an AD greater than 90. EIGRP will be used because its AD is better (lower) than the static route. If EIGRP goes down, however, the static route is used in its place. This is known as a *floating static route*.

Configuring an IPv4 Default Route

`Router(config)# ip route` `0.0.0.0 0.0.0.0 172.16.10.2`	Send all packets destined for networks not in my routing table to 172.16.10.2
`Router(config)# ip route` `0.0.0.0 0.0.0.0 serial 0/0/0`	Send all packets destined for networks not in my routing table out my serial 0/0/0 interface

NOTE: The combination of the 0.0.0.0 network address and the 0.0.0.0 mask is called a *quad-zero route*.

Verifying IPv4 Static Routes

To display the contents of the IP routing table, enter the following command:

`Router# show ip route`

NOTE: The codes to the left of the routes in the table tell you from where the router learned the routes. A static route is described by the letter *S*. A default route learned by a static route is described in the routing table by S*. The asterisk (*) indicates that the last path option is used when forwarding the packet.

`Router# show ip route`	Displays the current IPv4 routing table
`Router# show ip route` `summary`	Displays a summarized form of the current IPv4 routing table
`Router# show ip static route`	Displays only static IPv4 routes installed in the routing table
`Router# show ip static route` `172.16.10.0/24`	Displays only static route information about the specific address given
`Router# show ip static route` `172.16.10.0 255.255.255.0`	Displays only static route information about the specific address given
`Router# show ip static route` `summary`	Displays a summarized form of IPv4 static routes

Configuration Example: IPv4 Static Routes

Figure 15-1 illustrates the network topology for the configuration that follows, which shows how to configure static routes using the commands covered in this chapter.

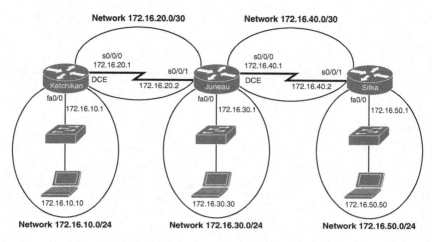

Figure 15-1   Network Topology for Static Route Configuration

NOTE: The host names, passwords, and interfaces have all been configured using the commands shown in the configuration example in Chapter 14, "Configuring a Cisco Router."

Ketchikan Router

`Ketchikan> `**`enable`**	Moves to privileged EXEC mode
`Ketchikan# `**`configure terminal`**	Moves to global configuration mode
`Ketchikan(config)# `**`ip route`**`172.16.30.0 255.255.255.0 172.16.20.2`	Configures a static route using the next-hop address. Since this is a point-to-point serial link, an exit interface should be used. This method works, but is inefficient
`Ketchikan(config)# `**`ip route`**`172.16.40.0 255.255.255.0 172.16.20.2`	Configures a static route using the next-hop address. Since this is a point-to-point serial link, an exit interface should be used. This method works, but is inefficient
`Ketchikan(config)# `**`ip route`**`172.16.50.0 255.255.255.0 172.16.20.2`	Configures a static route using the next-hop address. Since this is a point-to-point serial link, an exit interface should be used. This method works, but is inefficient
`Ketchikan(config)# `**`exit`**	Moves to privileged EXEC mode
`Ketchikan# `**`copy running-config startup-config`**	Saves the configuration to NVRAM

Juneau Router

`Juneau> `**`enable`**	Moves to privileged EXEC mode
`Juneau# `**`configure terminal`**	Moves to global configuration mode
`Juneau(config)# `**`ip route 172.16.10.0`**`255.255.255.0 `**`serial 0/0/1`**	Configures a static route using the exit interface

Juneau(config)# **ip route 172.16.50.0 255.255.255.0 serial 0/0/0**	Configures a static route using the exit interface
Juneau(config)# **exit**	Moves to privileged EXEC mode
Juneau# **copy running-config startup-config**	Saves the configuration to NVRAM

Sitka Router

Sitka> **enable**	Moves to privileged EXEC mode
Sitka# **configure terminal**	Moves to global configuration mode
Sitka(config)# **ip route 0.0.0.0 0.0.0.0 serial 0/0/1**	Configures a static route using the default route method. Note that an exit interface is used since this is a point-to-point link
Sitka(config)# **exit**	Moves to privileged EXEC mode
Sitka# **copy running-config startup-config**	Saves the configuration to NVRAM

Configuring an IPv6 Static Route

NOTE: To create a static route in IPv6, you use the same format as creating a static route in IPv4.

Figure 15-2 illustrates the network topology for the configuration that follows, which shows how to configure static routes with IPv6. Note that only the static routes on the Austin router are displayed.

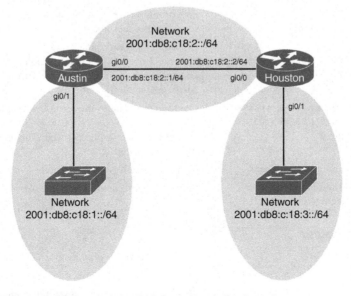

Figure 15-2 Network Topology for IPv6 Static Route Configuration

Austin(config)# **ipv6 route 2001:db8:c18:3::/64 2001:db8:c18:2::2**	Creates a static route configured to send all packets addressed to 2001:db8:c18:3::/64 to a next-hop address of 2001:db8:c18:2::2
Austin(config)# **ipv6 route 2001:db8:c18:3::/64 gigabitethernet 0/0**	Creates a directly attached static route configured to send packets out interface gigabitethernet 0/0
Austin(config)# **ipv6 route 2001:db8:c18:3::/64 gigabitethernet 0/0 2001:db8:c18:2::2**	Creates a fully specified static route on a broadcast interface. On a broadcast medium such as Ethernet, this is the preferred method for creating static routes

Floating Static Routes in IPv6

To create a static route with an AD set to 200 as opposed to the default AD of 1, enter the following command, for example:

Austin(config)# **ipv6 route 2001:db8:c18:3::/64 gigabitethernet 0/0 200**

NOTE: The default ADs used in IPv4 are the same for IPv6, with one exception—the AD number for unknown or unbelievable is 254 as opposed to 255. See Table 15-1 for a list of the default ADs.

Configuring an IPv6 Default Route

NOTE: To create a default route in IPv6, you use the same format as creating a default route in IPv4.

Austin(config)# **ipv6 route ::/0 2001:db8:c18:2::2**	Creates a default route configured to send all packets to a next-hop address of 2001:db8:c18:2::2
Austin(config)# **ipv6 route ::/0 serial 0/0/0**	Creates a default route configured to send packets out interface serial 0/0/0

Verifying IPv6 Static Routes

Router# **show ipv6 route**	Displays the current IPv6 routing table
Router# **show ipv6 route summary**	Displays a summarized form of the current IPv6 routing table
Router# **show ipv6 static**	Displays only static IPv6 routes installed in the routing table
Router# **show ipv6 static 2001:db8:5555:0/16**	Displays only static route information about the specific address given
Router# **show ipv6 static interface serial 0/0/0**	Displays only static route information with the specified interface as the outgoing interface
Router# **show ipv6 static detail**	Displays a more detailed entry for IPv6 static routes

CHAPTER 16

Open Shortest Path First (OSPF)

NOTE: The configuration and verification of OSPFv3 and multiarea OSPF in both OSPFv2 and OSPFv3 are not part of the blueprint for the new edition of the CCNA certification exam (200-301). Troubleshooting OSPF (v2 and v3) is part of the Implementing Cisco Enterprise Advanced Routing and Services exam (300-410), an optional concentration exam of the new CCNP Enterprise certification. Configuring multiarea OSPF is part of the Implementing Cisco Enterprise Network Core Technologies exam (300-401). Both of these CCNP exam topics will be in the new *CCNP Portable Command Guide*, to be released in early 2020.

This chapter provides information about the following topics:

- OSPFv2 versus OSPFv3
- Configuring OSPF
- Using wildcard masks with OSPF areas
- Loopback interfaces
- Router ID
- DR/BDR elections
- Timers
- Verifying OSPFv2 configurations
- Troubleshooting OSPFv2
- Configuration example: single-area OSPF

OSPFv2 Versus OSPFv3

The current version of Open Shortest Path First (OSPF), OSPFv2, was developed back in the late 1980s, when some parts of OSPF were designed to compensate for the inefficiencies of routers at that time. Now that router technology has dramatically improved, and with the arrival of IPv6, rather than modify OSPFv2 for IPv6, it was decided to create a new version of OSPF (OSPFv3), not just for IPv6, but for other newer technologies, too.

In most Cisco documentation, if you see something refer to OSPF, it is assumed to be referring to OSPFv2, and working with the IPv4 protocol stack.

The earliest release of the OSPFv3 protocol worked with IPv6 exclusively; if you needed to run OSPF for both IPv4 and IPv6, you had to have OSPFv2 and OSPFv3 running concurrently. Newer updates to OSPFv3 are allowing for OSPFv3 to handle both IPv4 and IPv6 addressing. The combining of IPv4 and IPv6 into OSPFv3 is not part of the

CCNA certification; it is part of the CCNP Enterprise certification and therefore out of scope for this guide. This guide works with the understanding that anything related to IPv4 will be using OSPFv2.

Configuring OSPF

`Router(config)# router ospf 123`	Starts OSPF process 123. The process ID is any positive integer value between 1 and 65,535. The process ID is not related to the OSPF area. The process ID merely distinguishes one process from another within the device
`Router(config-router)# network 172.16.10.0 0.0.0.255 area 0`	OSPF advertises interfaces, not networks. It uses the wildcard mask to determine which interfaces to advertise. Read this line to say, "Any interface with an address of 172.16.10.x is to run OSPF and be put into area 0" **NOTE:** The process ID number of one router does not have to match the process ID of any other router. Unlike Enhanced Interior Gateway Routing Protocol (EIGRP), matching this number across all routers does not ensure that network adjacencies will form
`Router(config-router)# log-adjacency-changes detail`	Configures the router to send a syslog message when there is a change of state between OSPF neighbors **TIP:** Although the **log-adjacency-changes** command is on by default, only up/down events are reported unless you use the **detail** keyword

CAUTION: Running two different OSPF processes does not create multiarea OSPF; it merely creates two separate instances of OSPF that do not communicate with each other. To create multiarea OSPF, you use two separate **network** statements and advertise two different links into different areas. Remember that multiarea OSPF is not part of the CCNA (200-301) vendor exam topics.

NOTE: You can enable OSPF directly on an interface with the **ip ospf** *process ID* **area** *area number* command. Because this command is configured directly on the interface, it takes precedence over the **network area** command entered in router configuration mode.

Using Wildcard Masks with OSPF Areas

When compared to an IP address, a wildcard mask identifies what addresses are matched to run OSPF and to be placed into an area:

- A 0 (zero) in a wildcard mask means to check the corresponding bit in the address for an exact match.
- A 1 (one) in a wildcard mask means to ignore the corresponding bit in the address—can be either 1 or 0.

Example 1: 172.16.0.0 0.0.255.255

172.16.0.0 = 10101100.00010000.00000000.00000000

0.0.255.255 = 00000000.00000000.11111111.11111111

Result = 10101100.00010000.*xxxxxxxx.xxxxxxxx*

172.16.*x.x* (Anything between 172.16.0.0 and 172.16.255.255 matches the example statement)

TIP: An octet in the wildcard mask of all 0s means that the octet has to match the address exactly. An octet in the wildcard mask of all 1s means that the octet can be ignored.

Example 2: 172.16.8.0 0.0.7.255

172.16.8.0 = 10101100.00010000.00001000.00000000

0.0.0.7.255 = 00000000.00000000.00000111.11111111

Result = 10101100.00010000.00001*xxx.xxxxxxxx*

00001*xxx* = 00001**000** to 00001**111** = 8–15

xxxxxxxx = 00000000 to 11111111 = 0–255

Anything between 172.16.8.0 and 172.16.15.255 matches the example statement

`Router(config-router)# network 172.16.10.1 0.0.0.0 area 0`	Read this line to say, "Any interface with an exact address of 172.16.10.1 is to run OSPF and be put into area 0"
`Router(config-router)# network 172.16.0.0 0.0.255.255 area 0`	Read this line to say, "Any interface with an address of 172.16.*x.x* is to run OSPF and be put into area 0"
`Router(config-router)# network 0.0.0.0 255.255.255.255 area 0`	Read this line to say, "Any interface with any address is to run OSPF and be put into area 0"

TIP: If you have problems determining which wildcard mask to use to place your interfaces into an OSPF area, use the **ip ospf** *process ID* **area** *area number* command directly on the interface.

`Router(config)# interface fastethernet 0/0`	Moves to interface configuration mode
`Router(config-if)# ip ospf 1 area 51`	Places this interface into area 51 of OSPF process 1
`Router(config-if)# interface gigabitethernet 0/0`	Moves to interface configuration mode
`Router(config-if)# ip ospf 1 area 0`	Places this interface into area 0 of OSPF process 1

TIP: If you assign interfaces to OSPF areas without first using the **router ospf** *x* command, the router creates the router process for you, and it shows up in a **show running-config** output.

Loopback Interfaces

Loopback interfaces are always "up and up" and do not go down unless manually shut down. This makes loopback interfaces great for use as an OSPF router ID.

`Router(config)#` `interface loopback0`	Creates a virtual interface named Loopback 0 and then moves the router to interface configuration mode

`Router(config-if)# ip`
`address 192.168.100.1`
`255.255.255.255`
	Assigns the IP address to the interface

Router ID

`Router(config)#` `router ospf 1`	Starts OSPF process 1
`Router(config-router)#` `router-id 10.1.1.1`	Sets the router ID to 10.1.1.1. If this command is used on an OSPF router process that is already active (has neighbors), the new router ID is used at the next reload or at a manual OSPF process restart
`Router(config-router)#` `no router-id 10.1.1.1`	Removes the static router ID from the configuration. If this command is used on an OSPF router process that is already active (has neighbors), the old router ID behavior is used at the next reload or at a manual OSPF process restart

NOTE: To choose the router ID at the time of OSPF process initialization, the router uses the following criteria in this specific order:

1. Use the router ID specified in the **router-id** *w.x.y.z* command.
2. Use the highest IP address of all active loopback interfaces on the router.
3. Use the highest IP address among all active nonloopback interfaces.

NOTE: To have the manually configured router ID take effect, you must clear the OSPF routing process with the **clear ip ospf process** command.

NOTE: There is no IPv6 form of router ID. All router IDs are 32-bit numbers in the form of an IPv4 address. Even if a router is running IPv6 exclusively, the router ID is still in the form of an IPv4 address.

DR/BDR Elections

`Router(config)# interface fastethernet0/0`	Enters interface configuration mode
`Router(config-if)# ip ospf priority 50`	Changes the OSPF interface priority to 50 **NOTE:** The assigned priority can be between 0 and 255. A priority of 0 makes the router ineligible to become a designated router (DR) or backup designated router (BDR). The highest priority wins the election and becomes the DR; the second highest priority becomes the BDR. A priority of 255 guarantees at least a tie in the election—assuming another router is also set to 255. If all routers have the same priority, regardless of the priority number, they tie. Ties are broken by the highest router ID. The default priority setting is 1 **TIP:** Do not assign the same priority value to more than one router

Timers

`Router(config-if)# ip ospf hello-interval 20`	Changes the hello interval timer to 20 seconds
`Router(config-if)# ip ospf dead-interval 80`	Changes the dead interval timer to 80 seconds

CAUTION: Hello and dead interval timers must match between two routers for those routers to become neighbors.

NOTE: The default hello timer is 10 seconds on multiaccess and point-to-point segments. The default hello timer is 30 seconds on nonbroadcast multiaccess (NBMA) segments such as Frame Relay, X.25, and ATM.

NOTE: The default dead interval timer is 40 seconds on multiaccess and point-to-point segments. The default dead timer is 120 seconds on NBMA segments such as Frame Relay, X.25, and ATM.

NOTE: If you change the hello interval timer, the dead interval timer is automatically adjusted to four times the new hello interval timer.

Verifying OSPFv2 Configurations

`Router# show ip protocol`	Displays parameters for all routing protocols running on the router
`Router# show ip route`	Displays a complete IP routing table
`Router# show ip route ospf`	Displays the OSPF routes in the routing table
`Router# show ip ospf`	Displays basic information about OSPF routing processes

Router# **show ip ospf border-routers**	Displays border and boundary router information
Router# **show ip ospf database**	Displays the contents of the OSPF database
Router# **show ip ospf database summary**	Displays a summary of the OSPF database
Router# **show ip ospf interface**	Displays OSPF info as it relates to all interfaces
Router# **show ip ospf interface fastethernet0/0**	Displays OSPF information for interface fastethernet 0/0
Router# **show ip ospf neighbor**	Lists all OSPF neighbors and their states
Router# **show ip ospf neighbor detail**	Displays a detailed list of neighbors

Troubleshooting OSPFv2

Router# **clear ip route ***	Clears the entire routing table, forcing it to rebuild
Router# **clear ip route a.b.c.d**	Clears a specific route to network a.b.c.d
Router# **clear ip ospf counters**	Resets OSPF counters
Router# **clear ip ospf process**	Resets the *entire* OSPF process, forcing OSPF to re-create neighbors, the database, and the routing table
Router# **clear ip ospf 3 process**	Resets OSPF process 3, forcing OSPF to re-create neighbors, the database, and the routing table
Router# **debug ip ospf events**	Displays all OSPF events
Router# **debug ip ospf adj**	Displays various OSPF states and DR/BDR election between adjacent routers
Router# **debug ip ospf packets**	Displays OSPF packets
Router# **undebug all**	Turns off all **debug** commands

Configuration Example: Single-Area OSPF

Figure 16-1 shows the network topology for the configuration that follows, which demonstrates how to configure single-area OSPF using the commands covered in this chapter.

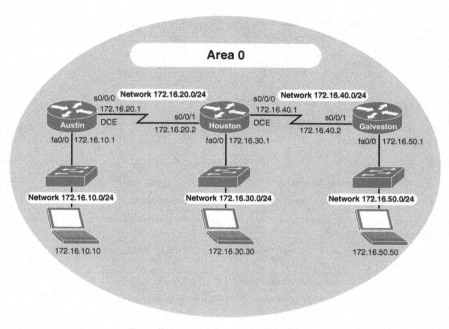

Figure 16-1   Network Topology for Single-Area OSPF Configuration

Austin Router

Router> **enable**	Moves to privileged EXEC mode
Router# **configure terminal**	Moves to global configuration mode
Router(config)# **hostname Austin**	Sets the host name
Austin(config)# **interface fastethernet 0/0**	Moves to interface configuration mode
Austin(config-if)# **ip address 172.16.10.1 255.255.255.0**	Assigns an IP address and a netmask
Austin(config-if)# **no shutdown**	Enables the interface
Austin(config-if)# **interface serial 0/0/0**	Moves to interface configuration mode
Austin(config-if)# **ip address 172.16.20.1 255.255.255.0**	Assigns an IP address and netmask
Austin(config-if)# **clock rate 2000000**	DCE cable plugged in this side
Austin(config-if)# **no shutdown**	Enables the interface
Austin(config-if)# **exit**	Returns to global configuration mode
Austin(config)# **router ospf 1**	Starts OSPF process 1
Austin(config-router)# **network 172.16.10.0 0.0.0.255 area 0**	Read this line to say, "Any interface with an address of 172.16.10.x is to run OSPF and be put into area 0"

`Austin(config-router)# network 172.16.20.0 0.0.0.255 area 0`	Read this line to say, "Any interface with an address of 172.16.20.x is to run OSPF and be put into area 0"
`Austin(config-router)#` Ctrl-Z	Returns to privileged EXEC mode
`Austin# copy running-config startup-config`	Saves the configuration to NVRAM

Houston Router

`Router> enable`	Moves to privileged EXEC mode
`Router# configure terminal`	Moves to global configuration mode
`Router(config)# hostname Houston`	Sets the host name
`Houston(config)# interface fastethernet 0/0`	Moves to interface configuration mode
`Houston(config-if)# ip address 172.16.30.1 255.255.255.0`	Assigns an IP address and netmask
`Houston(config-if)# no shutdown`	Enables the interface
`Houston(config-if)# interface serial0/0/0`	Moves to interface configuration mode
`Houston(config-if)# ip address 172.16.40.1 255.255.255.0`	Assigns an IP address and netmask
`Houston(config-if)# clock rate 2000000`	DCE cable plugged in this side
`Houston(config-if)# no shutdown`	Enables the interface
`Houston(config)# interface serial 0/0/1`	Moves to interface configuration mode
`Houston(config-if)# ip address 172.16.20.2 255.255.255.0`	Assigns an IP address and netmask
`Houston(config-if)# no shutdown`	Enables the interface
`Houston(config-if)# exit`	Returns to global configuration mode
`Houston(config)# router ospf 1`	Starts OSPF process 1
`Houston(config-router)# network 172.16.0.0 0.0.255.255 area 0`	Read this line to say, "Any interface with an address of 172.16.x.x is to run OSPF and be put into area 0" One statement now enables OSPF on all three interfaces
`Houston(config-router)#` Ctrl-Z	Returns to privileged EXEC mode
`Houston# copy running-config startup-config`	Saves the configuration to NVRAM

Galveston Router

`Router>` **`enable`**	Moves to privileged EXEC mode
`Router#` **`configure terminal`**	Moves to global configuration mode
`Router(config)#` **`hostname Galveston`**	Sets the host name
`Galveston(config)#` **`interface fastethernet 0/0`**	Moves to interface configuration mode
`Galveston(config-if)#` **`ip address 172.16.50.1 255.255.255.0`**	Assigns an IP address and netmask
`Galveston(config-if)#` **`no shutdown`**	Enables the interface
`Galveston(config-if)#` **`interface serial 0/0/1`**	Moves to interface configuration mode
`Galveston(config-if)#` **`ip address 172.16.40.2 255.255.255.0`**	Assigns an IP address and netmask
`Galveston(config-if)#` **`no shutdown`**	Enables the interface
`Galveston(config-if)#` **`exit`**	Returns to global configuration mode
`Galveston(config)#` **`router ospf 1`**	Starts OSPF process 1
`Galveston(config-router)#` **`network 172.16.40.2 0.0.0.0 area 0`**	Any interface with an exact address of 172.16.40.2 is to run OSPF and be put into area 0. This is the most precise way to place an exact address into the OSPF routing process
`Galveston(config-router)#` **`network 172.16.50.1 0.0.0.0 area 0`**	Read this line to say, "Any interface with an exact address of 172.16.50.1 is to be put into area 0"
`Galveston(config-router)#` Ctrl-Z	Returns to privileged EXEC mode
`Galveston#` **`copy running-config startup-config`**	Saves the configuration to NVRAM

CHAPTER 17

DHCP

This chapter provides information and commands concerning the following topics:

- Configuring a DHCP server on an IOS router
- Using Cisco IP Phones with a DHCP server
- Verifying and troubleshooting DHCP configuration
- Configuring a DHCP helper address
- Configuring a DHCP client on a Cisco IOS Software Ethernet interface
- Configuration example: DHCP

Configuring a DHCP Server on an IOS Router

Router(config)# **ip dhcp pool INTERNAL**	Creates a DHCP pool named INTERNAL. The name can be anything of your choosing
Router(dhcp-config)# **network 172.16.10.0 255.255.255.0**	Defines the range of addresses to be leased
Router(dhcp-config)# **default-router 172.16.10.1**	Defines the address of the default router for the client One IP address is required; however you can specify up to eight IP addresses in the command line. These are listed in order of precedence
Router(dhcp-config)# **dns-server 172.16.10.10**	Defines the address of the Domain Name System (DNS) server for the client.
Router(dhcp-config)# **netbios-name-server 172.16.10.10**	Defines the address of the NetBIOS server for the client
Router(dhcp-config)# **domain-name fakedomainname.com**	Defines the domain name for the client
Router(dhcp-config)# **lease 14 12 23**	Defines the lease time to be 14 days, 12 hours, 23 minutes
Router(dhcp-config)# **lease infinite**	Sets the lease time to infinity; the default time is 1 day
Router(dhcp-config)# **exit**	Returns to global configuration mode

Router(config)# **ip dhcp excluded-address 172.16.10.1 172.16.10.10**	Specifies the range of addresses not to be leased out to clients
Router(config)# **service dhcp**	Enables the DHCP service and relay features on a Cisco IOS router
Router(config)# **no service dhcp**	Turns the DHCP service off. The DHCP service is on by default in Cisco IOS Software

Using Cisco IP Phones with a DHCP Server

NOTE: This section is not part of the CCNA vendor exam topics.

Enterprises with small branch offices that implement a VoIP solution may choose to implement Cisco Unified Communications Manager, often referred to as CallManager, at a central office to control Cisco IP Phones at small branch offices. This design allows for centralized call processing and reduces equipment and administration required (especially at the branch office).

Cisco IP Phones download their configuration from a TFTP server. When a Cisco IP Phone starts, if it does not have its IP address and TFTP server IP address preconfigured, it sends a request with option 150 or 66 to the DHCP server to obtain this information.

- DHCP option 150 provides the IP address of a list of TFTP servers.

- DHCP option 66 gives the IP address of a single TFTP server.

NOTE: Cisco IP Phones may also include DHCP option 3 in their requests, which sets a default route.

Router(dhcp-config)# **option 66 ip 10.1.1.250**	Provides the IP address of a TFTP server for option 66
Router(dhcp-config)# **option 150 ip 10.1.1.250**	Provides the name of a TFTP server for option 150
Router(dhcp-config)# **option 150 ip 10.1.1.250 10.1.1.251**	Provides the names of two TFTP servers for option 150
Router(dhcp-config)# **option 3 ip 10.1.1.1**	Sets the default route

Verifying and Troubleshooting DHCP Configuration

Router# **show ip dhcp binding**	Displays a list of all bindings created
Router# **show ip dhcp binding** *w.x.y.z*	Displays the bindings for a specific DHCP client with an IP address of *w.x.y.z*
Router# **clear ip dhcp binding** *a.b.c.d*	Clears an automatic address binding from the DHCP server database
Router# **clear ip dhcp binding ***	Clears all automatic DHCP bindings

Router# **show ip dhcp conflict**	Displays a list of all address conflicts that the DHCP server recorded
Router# **clear ip dhcp conflict** *a.b.c.d*	Clears an address conflict from the database
Router# **clear ip dhcp conflict ***	Clears conflicts for all addresses
Router# **show ip dhcp database**	Displays recent activity on the DHCP database
Router# **show ip dhcp pool**	Displays information about DHCP address pools
Router# **show ip dhcp pool name**	Displays information about the DHCP pool named *name*
Router# **show ip dhcp server statistics**	Displays a list of the number of messages sent and received by the DHCP server
Router# **clear ip dhcp server statistics**	Resets all DHCP server counters to 0
Router# **debug ip dhcp server** {**events** | **packet** | **linkage** | **class**}	Displays the DHCP process of addresses being leased and returned

Configuring a DHCP Helper Address

| Router(config)# **interface gigabitethernet 0/0** | Moves to interface configuration mode |
| Router(config-if)# **ip helper-address 172.16.20.2** | Forwards DHCP broadcast messages as unicast messages to this specific address instead of having them be dropped by the router |

NOTE: The **ip helper-address** command forwards broadcast packets as a unicast to eight different UDP ports by default:

- TFTP (port 69)
- DNS (port 53)
- Time service (port 37)
- NetBIOS name server (port 137)
- NetBIOS datagram server (port 138)
- Boot Protocol (BOOTP) client and server datagrams (ports 67 and 68)
- TACACS service (port 49)

If you want to close some of these ports, use the **no ip forward-protocol udp** *x* command at the global configuration prompt, where *x* is the port number you want to close. The following command stops the forwarding of broadcasts to port 49:

```
Router(config)# no ip forward-protocol udp 49
```

If you want to open other UDP ports, use the **ip forward-helper udp** *x* command, where *x* is the port number you want to open:

```
Router(config)# ip forward-protocol udp 517
```

Configuring a DHCP Client on a Cisco IOS Software Ethernet Interface

`Router(config)# interface gigabitethernet 0/0`	Moves to interface configuration mode
`Router(config-if)# ip address dhcp`	Specifies that the interface acquire an IP address through DHCP

Configuration Example: DHCP

Figure 17-1 illustrates the network topology for the configuration that follows, which shows how to configure DHCP services on a Cisco IOS router using the commands covered in this chapter.

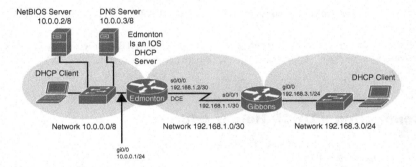

Figure 17-1   Network Topology for DHCP Configuration

Edmonton Router

`Router> enable`	Moves to privileged EXEC mode
`Router# configure terminal`	Moves to global configuration mode
`Router(config)# hostname Edmonton`	Sets the host name
`Edmonton(config)# interface gigabitethernet 0/0`	Moves to interface configuration mode
`Edmonton(config-if)# description LAN Interface`	Sets the local description of the interface
`Edmonton(config-if)# ip address 10.0.0.1 255.0.0.0`	Assigns an IP address and netmask
`Edmonton(config-if)# no shutdown`	Enables the interface
`Edmonton(config-if)# interface serial 0/0/0`	Moves to interface configuration mode
`Edmonton(config-if)# description Link to Gibbons Router`	Sets the local description of the interface
`Edmonton(config-if)# ip address 192.168.1.2 255.255.255.252`	Assigns an IP address and netmask

Edmonton(config-if)# **clock rate 2000000**	Assigns the clock rate to the DCE cable on this side of link
Edmonton(config-if)# **no shutdown**	Enables the interface
Edmonton(config-if)# **exit**	Returns to global configuration mode
Edmonton(config)# **ip route 192.168.3.0 255.255.255.0 serial 0/0/0**	Creates a static route to the destination network
Edmonton(config)# **service dhcp**	Verifies that the router can use DHCP services and that DHCP is enabled
Edmonton(config)# **ip dhcp pool 10NETWORK**	Creates a DHCP pool called 10NETWORK
Edmonton(dhcp-config)# **network 10.0.0.0 255.0.0.0**	Defines the range of addresses to be leased
Edmonton(dhcp-config)# **default-router 10.0.0.1**	Defines the address of the default router for clients
Edmonton(dhcp-config)# **netbios-name-server 10.0.0.2**	Defines the address of the NetBIOS server for clients
Edmonton(dhcp-config)# **dns-server 10.0.0.3**	Defines the address of the DNS server for clients
Edmonton(dhcp-config)# **domain-name fakedomainname.com**	Defines the domain name for clients
Edmonton(dhcp-config)# **lease 12 14 30**	Sets the lease time to be 12 days, 14 hours, 30 minutes
Edmonton(dhcp-config)# **exit**	Returns to global configuration mode
Edmonton(config)# **ip dhcp excluded-address 10.0.0.1 10.0.0.5**	Specifies the range of addresses not to be leased out to clients
Edmonton(config)# **ip dhcp pool 192.168.3NETWORK**	Creates a DHCP pool called the 192.168.3NETWORK
Edmonton(dhcp-config)# **network 192.168.3.0 255.255.255.0**	Defines the range of addresses to be leased
Edmonton(dhcp-config)# **default-router 192.168.3.1**	Defines the address of the default router for clients
Edmonton(dhcp-config)# **netbios-name-server 10.0.0.2**	Defines the address of the NetBIOS server for clients
Edmonton(dhcp-config)# **dns-server 10.0.0.3**	Defines the address of the DNS server for clients
Edmonton(dhcp-config)# **domain-name fakedomainname.com**	Defines the domain name for clients
Edmonton(dhcp-config)# **lease 12 14 30**	Sets the lease time to be 12 days, 14 hours, 30 minutes
Edmonton(dhcp-config)# **exit**	Returns to global configuration mode
Edmonton(config)# **exit**	Returns to privileged EXEC mode
Edmonton# **copy running-config startup-config**	Saves the configuration to NVRAM

Gibbons Router

`Router> enable`	Moves to privileged EXEC mode
`Router# configure terminal`	Moves to global configuration mode
`Router(config)# hostname Gibbons`	Sets the host name
`Gibbons(config)# interface gigabitethernet 0/0`	Moves to interface configuration mode
`Gibbons(config-if)# description LAN Interface`	Sets the local description of the interface
`Gibbons(config-if)# ip address 192.168.3.1 255.255.255.0`	Assigns an IP address and netmask
`Gibbons(config-if)# ip helper-address 192.168.1.2`	Forwards DHCP broadcast messages as unicast messages to this specific address instead of having them be dropped by the router
`Gibbons(config-if)# no shutdown`	Enables the interface
`Gibbons(config-if)# interface serial 0/0/1`	Moves to interface configuration mode
`Gibbons(config-if)# description Link to Edmonton Router`	Sets the local description of the interface
`Gibbons(config-if)# ip address 192.168.1.1 255.255.255.252`	Assigns an IP address and netmask
`Gibbons(config-if)# no shutdown`	Enables the interface
`Gibbons(config-if)# exit`	Returns to global configuration mode
`Gibbons(config)# ip route 0.0.0.0 0.0.0.0 serial 0/0/1`	Creates a default static route to the destination network
`Gibbons(config)# exit`	Returns to privileged EXEC mode
`Gibbons# copy running-config startup-config`	Saves the configuration to NVRAM

CHAPTER 18

Network Address Translation (NAT)

This chapter provides information and commands concerning the following topics:

- Private IP addresses: RFC 1918
- Configuring dynamic NAT: One private to one public address translation
- Configuring PAT: Many private to one public address translation
- Configuring static NAT: One private to one permanent public address translation
- Verifying NAT and PAT configurations
- Troubleshooting NAT and PAT configurations
- Configuration example: PAT

Private IP Addresses: RFC 1918

Table 18-1 lists the address ranges as specified in RFC 1918 that anyone can use as internal private addresses. These will be your "inside-the-LAN" addresses that will have to be translated into public addresses that can be routed across the Internet. Any network is allowed to use these addresses; however, these addresses are not allowed to be routed onto the public Internet.

TABLE 18-1   RFC 1918 Private Address Ranges

Internal Address Range	CIDR Prefix	Traditional Class
10.0.0.0–10.255.255.255	10.0.0.0/8	A
172.16.0.0–172.31.255.255	172.16.0.0/12	B
192.168.0.0–192.168.255.255	192.168.0.0/16	C

Configuring Dynamic NAT: One Private to One Public Address Translation

CAUTION: Make sure that you have in your router configurations a way for packets to travel back to your NAT router. Include on the ISP router a static route defining a path to your NAT addresses/networks and how to travel back to your internal network. Without this in place, a packet can leave your network with a public address, but it cannot return if your ISP router does not know where the public addresses exist in the network. You should be advertising the public addresses, not your private addresses.

Dynamic Address Translation (Dynamic NAT) maps unregistered (private) IP addresses to registered (public) IP addresses from a pool of registered IP addresses.

Figure 18-1 shows the network topology for the Dynamic NAT configuration that follows using the commands covered in this chapter.

Public Address Pool for NAT:
64.64.64.64/192

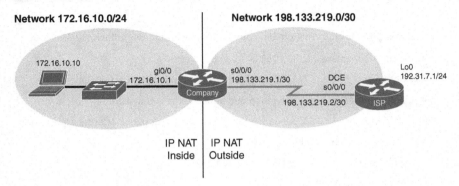

Figure 18-1 One Private to One Public Dynamic Address Translation Configuration

Step 1: Define a static route on the remote router stating where the public addresses should be routed.	`ISP(config)# ip route 64.64.64.64 255.255.255.192 s0/0/0`	Informs the ISP router where to send packets with addresses destined for 64.64.64.64 255.255.255.192
Step 2: Define a pool of usable public IP addresses on your local router that will perform NAT.	`Company(config)# ip nat pool scott 64.64.64.65 64.64.64.126 netmask 255.255.255.192`	The private address will receive the first available public address in the pool Defines the following: The name of the pool is *scott* (but can be anything) The start of the pool is 64.64.64.65 The end of the pool is 64.64.64.126 The subnet mask is 255.255.255.192
Step 3: Create an access control list (ACL) that will identify which private IP addresses will be translated.	`Company(config)# access-list 1 permit 172.16.10.0 0.0.0.255`	
Step 4: Link the ACL to the pool of addresses (create the translation).	`Company(config)# ip nat inside source list 1 pool scott`	Defines the following: The source of the private addresses is from ACL 1 The pool of available public addresses is named *scott*

Step 5: Define which interfaces are inside (contain the private addresses).	`Company(config)#` **`interface`** **`gigabitethernet 0/0`**	Moves to interface configuration mode
	`Company` `(config-if)#` **`ip nat inside`**	You can have more than one inside interface on a router. Addresses from each inside interface are then allowed to be translated into a public address
Step 6: Define the outside interface (the interface leading to the public network).	`Company` `(config-if)#` **`exit`**	Returns to global configuration mode
	`Company(config)#` **`interface serial`** **`0/0/0`**	Moves to interface configuration mode
	`Company` `(config-if)#` **`ip`** **`nat outside`**	Defines which interface is the outside interface for NAT

Configuring PAT: Many Private to One Public Address Translation

PAT maps multiple unregistered (private) IP addresses to a single registered (public) IP address (many to one) using different ports. This is also known as overloading or overload translations. By using PAT or overloading, thousands of users can be connected to the Internet by using only one real registered public IP address.

Figure 18-2 shows the network topology for the PAT configuration that follows using the commands covered in this chapter.

Public Address Pool Range for NAT: 64.64.64.65–70/192

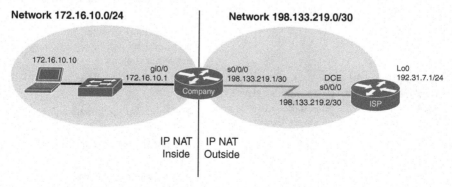

Figure 18-2   Many Private to One Public Address Translation Configuration

Step 1: Define a static route on the remote router stating where public addresses should be routed.	`ISP(config)# ip` `route 64.64.64.64` `255.255.255.192` `s0/0/0`	Informs the Internet service provider (ISP) router where to send packets with addresses destined for 64.64.64.64 255.255.255.192
Step 2: Define a pool of usable public IP addresses on your local router that will perform NAT (optional).	`Company(config)#` `ip nat pool` `scott 64.64.64.65` `64.64.64.70 netmask` `255.255.255.192`	Use this step if you have many private addresses to translate. A single public IP address can handle thousands of private addresses. Without using a pool of addresses, you can translate all private addresses into the IP address of the exit interface (the serial link to the ISP, for example) Defines the following: The name of the pool is *scott* (but can be anything) The start of the pool is 64.64.64.65 The end of the pool is 64.64.64.70 The subnet mask is 255.255.255.192
Step 3: Create an ACL that will identify which private IP addresses will be translated.	`Company(config)#` `access-list 1` `permit 172.16.10.0` `0.0.0.255`	
Step 4 (Option 1): Link the ACL to the outside public interface (create the translation).	`Company(config)#` `ip nat inside` `source list 1` `interface serial` `0/0/0 overload`	The source of the private addresses is from ACL 1 The public address to be translated into is the one assigned to serial 0/0/0 The **overload** keyword states that port numbers will be used to handle many translations
Step 4 (Option 2): Link the ACL to the pool of addresses (create the translation).	`Company(config)#` `ip nat inside` `source list 1 pool` `scott overload`	If using the pool created in Step 2 ... The source of the private addresses is from ACL 1 The pool of the available addresses is named *scott* The **overload** keyword states that port numbers will be used to handle many translations

Step 5: Define which interfaces are inside (contain the private addresses).	`Company(config)#`
	`interface`
	`gigabitethernet 0/0`
		Moves to interface configuration mode

	`Company(config-if)#` `ip nat inside`	You can have more than one inside interface on a router
Step 6: Define the outside interface (the interface leading to the public network).	`Company(config-if)#` `exit`	Returns to global configuration mode

	`Company(config)#`
	`interface serial`
	`0/0/0`
		Moves to interface configuration mode

	`Company(config-if)#` `ip nat outside`	Defines which interface is the outside interface for NAT

NOTE: You can have an IP NAT pool of more than one address, if needed. The syntax for this is as follows:

```
Corp(config)# ip nat pool scott 64.64.64.70 64.64.64.75 netmask
255.255.255.128
```

You would then have a pool of six addresses (and all their ports) available for translation.

NOTE: The theoretical maximum number of translations between internal addresses and a single outside address using PAT is 65,536. Port numbers are encoded in a 16-bit field, so 2^{16} = 65,536.

Configuring Static NAT: One Private to One Permanent Public Address Translation

Static Network Address Translation (Static NAT) allows one-to-one mapping between local (private) and global (public) IP addresses.

Figure 18-3 shows the network topology for the Static NAT configuration that follows using the commands covered in this chapter.

Static NAT Translations:
172.16.10.5 = 64.64.64.65
172.16.10.9 = 64.64.64.66
172.16.10.10 = 64.64.64.67

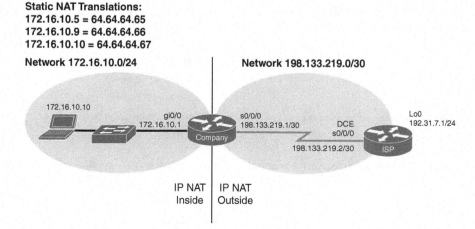

Figure 18-3   One Private to One Permanent Public Address Translation Configuration

Step 1: Define a static route on the remote router stating where the public addresses should be routed.	`ISP(config)# ip` `route 64.64.64.64` `255.255.255.192 s0/0/0`	Informs the ISP router where to send packets with addresses destined for 64.64.64.64 255.255.255.192
Step 2: Create a static mapping on your local router that will perform NAT.	`Company(config)# ip nat` `inside source static` `172.16.10.5 64.64.64.65` `Company(config)# ip nat` `inside source static` `172.16.10.9 64.64.64.66` `Company(config)# ip nat` `inside source static` `172.16.10.10 64.64.64.67`	Permanently translates the inside address of 172.16.10.5 to a public address of 64.64.64.65 Use the command for each of the private IP addresses you want to statically map to a public address
Step 3: Define which interfaces are inside (contain the private addresses).	`Company(config)# interface` `gigabitethernet 0/0`	Moves to interface configuration mode
	`Company(config-if)# ip` `nat inside`	You can have more than one inside interface on a router
Step 4: Define the outside interface (the interface leading to the public network).	`Company(config-if)#` `interface serial 0/0/0`	Moves to interface configuration mode
	`Company(config-if)# ip` `nat outside`	Defines which interface is the outside interface for NAT

Verifying NAT and PAT Configurations

`Router# show access-list`	Displays access lists
`Router# show ip nat translations`	Displays the translation table
`Router# show ip nat statistics`	Displays NAT statistics
`Router# clear ip nat translation` `inside 1.1.1.1 2.2.2.2 outside` `3.3.3.3 4.4.4.4`	Clears a specific translation from the table before it times out 1.1.1.1 = Global IP address 2.2.2.2 = Local IP address 3.3.3.3 = Local IP address 4.4.4.4 = Global IP address
`Router# clear ip nat translation *`	Clears the entire translation table before entries time out

NOTE: The default timeout for a translation entry in a NAT table is 24 hours.

Troubleshooting NAT and PAT Configurations

Router# **debug ip nat**	Displays information about every packet that is translated
	Be careful with this command. The router's CPU might not be able to handle this amount of output and might therefore hang the system
Router# **debug ip nat detailed**	Displays greater detail about packets being translated

Configuration Example: PAT

Figure 18-4 shows the network topology for the PAT configuration that follows using the commands covered in this chapter.

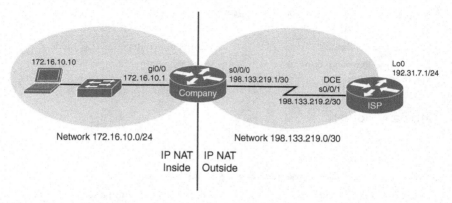

Figure 18-4 Port Address Translation Configuration

ISP Router

Router> **enable**	Moves to privileged EXEC mode
Router# **configure terminal**	Moves to global configuration mode
Router(config)# **hostname ISP**	Sets the host name
ISP(config)# **no ip domain-lookup**	Turns off Domain Name System (DNS) resolution to avoid wait time due to DNS lookup of spelling errors
ISP(config)# **enable secret cisco**	Sets the encrypted password to *cisco*
ISP(config)# **line console 0**	Moves to line console mode
ISP(config-line)# **login**	User must log in to be able to access the console port
ISP(config-line)# **password class**	Sets the console line password to *class*
ISP(config-line)# **logging synchronous**	Commands will be appended to a new line

`ISP(config-line)# exit`	Returns to global configuration mode
`ISP(config)# interface serial 0/0/1`	Moves to interface configuration mode
`ISP(config-if)# ip address 198.133.219.2 255.255.255.252`	Assigns an IP address and netmask
`ISP(config-if)# clock rate 56000`	Assigns the clock rate to the DCE cable on this side of the link
`ISP(config-if)# no shutdown`	Enables the interface
`ISP(config-if)# interface loopback 0`	Creates loopback interface 0 and moves to interface configuration mode
`ISP(config-if)# ip address 192.31.7.1 255.255.255.255`	Assigns an IP address and netmask
`ISP(config-if)# exit`	Returns to global configuration mode
`ISP(config)# ip route 64.64.64.64 255.255.255.192 s0/0/0`	Creates a static route so that packets can be sent to correct destination
`ISP(config)# exit`	Returns to privileged EXEC mode
`ISP# copy running-config startup-config`	Saves the configuration to NVRAM

Company Router

`Router> enable`	Moves to privileged EXEC mode
`Router# configure terminal`	Moves to global configuration mode
`Router(config)# hostname Company`	Sets the host name
`Company(config)# no ip domain-lookup`	Turns off DNS resolution to avoid wait time due to DNS lookup of spelling errors
`Company(config)# enable secret cisco`	Sets the secret password to *cisco*
`Company(config)# line console 0`	Moves to line console mode
`Company(config-line)# login`	User must log in to be able to access the console port
`Company(config-line)# password class`	Sets the console line password to *class*
`Company(config-line)# logging synchronous`	Commands will be appended to a new line
`Company(config-line)# exit`	Returns to global configuration mode
`Company(config)# interface gigabitethernet 0/0`	Moves to interface configuration mode
`Company(config-if)# ip address 172.16.10.1 255.255.255.0`	Assigns an IP address and netmask
`Company(config-if)# no shutdown`	Enables the interface
`Company(config-if)# interface serial 0/0/0`	Moves to interface configuration mode

`Company(config-if)# ip address 198.133.219.1 255.255.255.252`	Assigns an IP address and netmask
`Company(config-if)# no shutdown`	Enables the interface
`Company(config-if)# exit`	Returns to global configuration mode
`Company(config)# ip route 0.0.0.0 0.0.0.0 198.133.219.2`	Sends all packets not defined in the routing table to the ISP router
`Company(config)# access-list 1 permit 172.16.10.0 0.0.0.255`	Defines which addresses are permitted through; these addresses are those that will be allowed to be translated with NAT
`Company(config)# ip nat inside source list 1 interface serial 0/0/0 overload`	Creates NAT by combining list 1 with the interface serial 0/0/0. Overloading will take place
`Company(config)# interface gigabitethernet 0/0`	Moves to interface configuration mode
`Company(config-if)# ip nat inside`	Location of private inside addresses
`Company(config-if)# interface serial 0/0/0`	Moves to interface configuration mode
`Company(config-if)# ip nat outside`	Location of public outside addresses
`Company(config-if)# Ctrl-Z`	Returns to privileged EXEC mode
`Company# copy running-config startup-config`	Saves the configuration to NVRAM

CHAPTER 19

Configuring Network Time Protocol (NTP)

This chapter provides information about the following topics:

- NTP configuration
- NTP design
- Securing NTP
- Verifying and troubleshooting NTP
- Setting the clock on a router
- Using time stamps
- Configuration example: NTP

Most networks today are being designed with high performance and reliability in mind. Delivery of content is, in many cases, guaranteed by service level agreements (SLAs). Having your network display an accurate time is vital to ensuring that you have the best information possible when reading logging messages or troubleshooting issues.

NTP Configuration

`Edmonton(config)# ntp server` `209.165.200.254`	Configures the Edmonton router to synchronize its clock to a public NTP server at address 209.165.200.254 **NOTE:** This command makes the Edmonton router an NTP client to the external NTP server **NOTE:** A Cisco IOS router can be both a client to an external NTP server and an NTP server to client devices inside its own internal network **NOTE:** When NTP is enabled on a Cisco IOS router, it is enabled on all interfaces
`Edmonton(config)# ntp server` `209.165.200.234 prefer`	Specifies a preferred NTP server if multiple ones are configured **TIP:** It is recommended to configure more than one NTP server
`Edmonton(config-if)#` `ntp disable`	Disables the NTP server function on a specific interface. The interface will still act as an NTP client **TIP:** Use this command on interfaces connected to external networks

`Edmonton(config)# ntp master` `stratum`	Configures the router to be an NTP master clock to which peers synchronize when no external NTP source is available. The *stratum* is an optional number between 1 and 15. When enabled, the default stratum is 8 **NOTE:** A reference clock (for example, an atomic clock) is said to be a stratum-0 device. A stratum-1 server is directly connected to a stratum-0 device. A stratum-2 server is connected across a network path to a stratum-1 server. The larger the stratum number (moving toward 15), the less authoritative that server is and the less accuracy it will have
`Edmonton(config)#` `ntp max-associations 200`	Configures the maximum number of NTP peer-and-client associations that the router will serve. The range is 0 to 4,294,967,295. The default is 100

`Edmonton(config)# access`
`list 101 permit udp any host`
`a.b.c.d eq ntp`
	Creates an access list statement that will allow NTP communication for the NTP server at address *a.b.c.d*. This ACL should be placed in an inbound direction

NOTE: When a local device is configured with the **ntp master** command, it can be identified by a syntactically correct but invalid IP address. This address will be in the form of 127.127.*x.x*. The master will synchronize with itself and uses the 127.127.*x.x* address to identify itself. This address will be displayed with the **show ntp associations** command and must be permitted via an access list if you are authenticating your NTP servers.

NTP Design

You have two different options in NTP design: flat and hierarchical. In a flat design, all routers are peers to each other. Each router is both a client and a server with every other router. In a hierarchical model, there is a preferred order of routers that are servers and others that act as clients. You use the **ntp peer** command to determine the hierarchy.

TIP: Do not use the flat model in a large network, because with many NTP servers it can take a long time to synchronize the time.

`Edmonton(config)#` `ntp peer 172.16.21.1`	Configures an IOS device to synchronize its software clock to a peer at 172.16.21.1

`Edmonton(config)#`
`ntp peer 172.16.21.1`
`version 2`
	Configures an IOS device to synchronize its software clock to a peer at 172.16.21.1 using version 2 of NTP. There are three versions of NTP (versions 2–4)

NOTE: Although Cisco IOS recognizes three versions of NTP, versions 3 and 4 are most commonly used. Version 4 introduces support for IPv6 and is backward compatible with version 3. NTPv4 also adds DNS support for IPv6.

NOTE: NTPv4 has increased security support using public key cryptography and X.509 certificates.

NOTE: NTPv3 uses broadcast messages. NTPv4 uses multicast messages.

Edmonton(config)# **ntp peer 172.16.21.1** **source loopback 0**	Configures an IOS device to synchronize its software clock to a peer at 172.16.21.1. The source IP address is the address of interface Loopback 0 **TIP:** Choose a loopback interface as your source for NTP because it will never go down. ACL statements will also be easier to write as you will require only one line to allow or deny traffic
Edmonton(config)# **ntp peer 172.16.21.1** **source loopback 0 prefer**	Makes this peer the preferred peer that provides synchronization

Securing NTP

You can secure NTP operation using authentication and access lists.

NOTE: Securing NTP is not part of the CCNA (200-301) exam topics.

Enabling NTP Authentication

NTPServer(config)# **ntp authentication-key 1** **md5 NTPpa55word**	Defines an NTP authentication key **1** = number of authentication key. Can be a number between 1 and 4,294,967,295 **md5** = using MD5 hash. This is the only option available on Cisco devices **NTPpa55word** = password associated with this key
NTPServer(config)# **ntp authenticate**	Enables NTP authentication
NTPServer(config)# **ntp trusted-key 1**	Defines which keys are valid for NTP authentication. The key number here must match the key number you defined in the **ntp authentication-key** command
NTPClient(config)# **ntp authentication-key** **1 md5 NTPpa55word**	Defines an NTP authentication key
NTPClient(config)# **ntp authenticate**	Enables NTP authentication
NTPClient(config)# **ntp trusted-key 1**	Defines which keys are valid for NTP authentication. The key number here must match the key number you defined in the **ntp authentication-key** command
NTPClient(config)# **ntp** **server 192.168.200.1** **key 1**	Defines the NTP server that requires authentication at address 192.168.200.1 and identifies the peer key number as key 1

NOTE: NTP does not authenticate clients; it only authenticates the source. That means that a device will respond to unauthenticated requests. Therefore, access lists should be used to limit NTP access.

NOTE: Once a device is synchronized to an NTP source, it will become an NTP server to any device that requests synchronization.

Limiting NTP Access with Access Lists

Edmonton(config)# **access-list 1 permit** **10.1.0.0 0.0.255.255**	Defines an access list that permits only packets with a source address of 10.1.*x.x*
Edmonton(config)# **ntp access-group** **peer 1**	Creates an access group to control NTP access and applies access list 1. The **peer** keyword enables the device to receive time requests and NTP control queries and to synchronize itself to servers specified in the access list
Edmonton(config)# **ntp access-group** **serve 1**	Creates an access group to control NTP access and applies access list 1. The **serve** keyword enables the device to receive time requests and NTP control queries from the servers specified in the access list but not to synchronize itself to the specified servers
Edmonton(config)# **ntp access-group** **serve-only 1**	Creates an access group to control NTP access and applies access list 1. The **serve-only** keyword enables the device to receive only time requests from servers specified in the access list
Edmonton(config)# **ntp access-group** **query-only 1**	Creates an access group to control NTP access and applies access list 1. The **query-only** keyword enables the device to receive only NTP control queries from the servers specified in the access list

NOTE: NTP access group options are scanned from least restrictive to most restrictive in the following order: **peer, serve, serve-only, query-only**. However, if NTP matches a deny ACL rule in a configured peer, ACL processing stops and does not continue to the next access group option.

Verifying and Troubleshooting NTP

Edmonton# **show ntp associations**	Displays the status of NTP associations
Edmonton# **show ntp associations detail**	Displays detailed information about each NTP association
Edmonton# **show ntp status**	Displays the status of the NTP. This command shows whether the router's clock has synchronized with the external NTP server
Edmonton# **debug ip packets**	Checks to see whether NTP packets are received and sent
Edmonton# **debug ip packet 1**	Limits debug output to ACL 1
Edmonton# **debug ntp adjust**	Displays debug output for NTP clock adjustments
Edmonton# **debug ntp all**	Displays all NTP debugging output
Edmonton# **debug ntp events**	Displays all NTP debugging events
Edmonton# **debug ntp packet**	Displays NTP packet debugging; lets you see the time that the peer/server gives you in a received packet
Edmonton# **debug ntp packet detail**	Displays detailed NTP packet dump
Edmonton# **debug ntp packet peer** *a.b.c.d*	Displays debugging from NTP peer at address *a.b.c.d*

Setting the Clock on a Router

NOTE: It is important to have your routers display the correct time for use with time stamps and other logging features.

If the system is synchronized by a valid outside timing mechanism, such as an NTP server, or if you have a router with a hardware clock, you do not need to set the software clock. Use the software clock if no other time sources are available.

Edmonton# `calendar set` `16:30:00 23 June 2019`	Manually sets the system hardware clock. Time is set using military (24-hour) format. The hardware clock runs continuously, even if the router is powered off or rebooted
Edmonton# `show calendar`	Displays the hardware calendar
Edmonton(config)# `clock` `calendar-valid`	Configures the system as an authoritative time source for a network based on its hardware clock **NOTE:** Because the hardware clock is not as accurate as other time sources (it runs off of a battery), you should use this only when a more accurate time source (such as NTP) is not available
Edmonton# `clock` `read-calendar`	Manually reads the hardware clock settings into the software clock
Edmonton# `clock set` `16:30:00 23 June 2019`	Manually sets the system software clock. Time is set using military (24-hour) format
Edmonton(config)# `clock summer-time` *zone* `recurring` [*week day* *month hh:mm week day* *month hh:mm* [*offset*]]	Configures the system to automatically switch to summer time (daylight saving time) **NOTE:** Summer time is disabled by default Arguments for the command are as follows: *zone*: Name of the time zone
Edmonton(config)# `clock` `summer-time` *zone* `date` *date month year hh:mm* *date month year hh:mm* [*offset*]	**recurring:** Summer time should start and end on the corresponding specified days every year **date:** Indicates that summer time should start on the first specific date listed in the command and end on the second specific date in the command *week*: (Optional) Week of the month (1 to 5 or last)
Edmonton(config)# `clock` `summer-time` *zone* `date` *month date year hh:mm* *month date year hh:mm* [*offset*]	*day*: (Optional) Day of the week (Sunday, Monday, and so on) *date*: Date of the month (1 to 31) *month*: (Optional) Month (January, February, and so on) *year*: Year (1993 to 2035) *hh:mm*: (Optional) Time (military format) in hours and minutes *offset*: (Optional) Number of minutes to add during summer time (default is 60)

Edmonton(config)# **clock timezone** *zone* *hours-offset* [*minutes-offset*]	Configures the time zone for display purposes. To set the time to Coordinated Universal Time (UTC), use the no form of this command *zone*: Name of the time zone to be displayed when standard time is in effect. See Tables 19-1 and 19-2 for common time zone acronyms *hours-offset*: Hours difference from UTC *minutes-offset*: (Optional) Minutes difference from UTC
Edmonton(config)# **clock timezone PST -8**	Configures the time zone to Pacific Standard Time, which is 8 hours behind UTC
Edmonton(config)# **clock timezone NL -3 30**	Configures the time zone to Newfoundland Time for Newfoundland, Canada, which is 3.5 hours behind UTC
Edmonton# **clock update-calendar**	Updates the hardware clock from the software clock
Edmonton# **show clock**	Displays the time and date from the system software clock
Edmonton# **show clock detail**	Displays the clock source (NTP, hardware) and the current summer-time setting (if any)

Table 19-1 shows the common acronyms used for setting the time zone on a router.

TABLE 19-1   Common Time Zone Acronyms

Region/Acronym	Time Zone Name and UTC Offset
Europe	
GMT	Greenwich Mean Time, as UTC
BST	British Summer Time, as UTC +1 hour
IST	Irish Summer Time, as UTC +1 hour
WET	Western Europe Time, as UTC
WEST	Western Europe Summer Time, as UTC +1 hour
CET	Central Europe Time, as UTC +1
CEST	Central Europe Summer Time, as UTC +2
EET	Eastern Europe Time, as UTC +2
EEST	Eastern Europe Summer Time, as UTC +3
MSK	Moscow Time, as UTC +3
MSD	Moscow Summer Time, as UTC +4
United States and Canada	
AST	Atlantic Standard Time, as UTC –4 hours
ADT	Atlantic Daylight Time, as UTC –3 hours
ET	Eastern Time, either as EST or EDT, depending on place and time of year
EST	Eastern Standard Time, as UTC –5 hours
EDT	Eastern Daylight Time, as UTC –4 hours

Region/Acronym	Time Zone Name and UTC Offset
CT	Central Time, either as CST or CDT, depending on place and time of year
CST	Central Standard Time, as UTC –6 hours
CDT	Central Daylight Time, as UTC –5 hours
MT	Mountain Time, either as MST or MDT, depending on place and time of year
MST	Mountain Standard Time, as UTC –7 hours
MDT	Mountain Daylight Time, as UTC –6 hours
PT	Pacific Time, either as PST or PDT, depending on place and time of year
PST	Pacific Standard Time, as UTC –8 hours
PDT	Pacific Daylight Time, as UTC –7 hours
AKST	Alaska Standard Time, as UTC –9 hours
AKDT	Alaska Standard Daylight Time, as UTC –8 hours
HST	Hawaiian Standard Time, as UTC –10 hours
Australia	
WST	Western Standard Time, as UTC +8 hours
CST	Central Standard Time, as UTC +9.5 hours
EST	Eastern Standard/Summer Time, as UTC +10 hours (+11 hours during summer time)

Table 19-2 lists an alternative method for referring to time zones, in which single letters are used to refer to the time zone difference from UTC. Using this method, the letter *Z* is used to indicate the zero meridian, equivalent to UTC, and the letter *J* (Juliet) is used to refer to the local time zone. Using this method, the international date line is between time zones M and Y.

TABLE 19-2   Single-Letter Time Zone Designators

Letter Designator	Word Designator	Difference from UTC
Y	Yankee	UTC –12 hours
X	X-ray	UTC –11 hours
W	Whiskey	UTC –10 hours
V	Victor	UTC –9 hours
U	Uniform	UTC –8 hours
T	Tango	UTC –7 hours
S	Sierra	UTC –6 hours
R	Romeo	UTC –5 hours
Q	Quebec	UTC –4 hours

P	Papa	UTC −3 hours
O	Oscar	UTC −2 hours
N	November	UTC −1 hour
Z	Zulu	Same as UTC
A	Alpha	UTC +1 hour
B	Bravo	UTC +2 hours
C	Charlie	UTC +3 hours
D	Delta	UTC +4 hours
E	Echo	UTC +5 hours
F	Foxtrot	UTC +6 hours
G	Golf	UTC +7 hours
H	Hotel	UTC +8 hours
I	India	UTC +9 hours
K	Kilo	UTC +10 hours
L	Lima	UTC +11 hours
M	Mike	UTC +12 hours

Using Time Stamps

Edmonton(config)# **service timestamps**	Adds a time stamp to all system logging messages
Edmonton(config)# **service timestamps debug**	Adds a time stamp to all debugging messages
Edmonton(config)# **service timestamps debug uptime**	Adds a time stamp along with the total uptime of the router to all debugging messages
Edmonton(config)# **service timestamps debug datetime localtime**	Adds a time stamp displaying the local time and the date to all debugging messages
Edmonton(config)# **no service timestamps**	Disables all time stamps

Configuration Example: NTP

Figure 19-1 shows the network topology for the configuration that follows, which demonstrates how to configure NTP using the commands covered in this chapter.

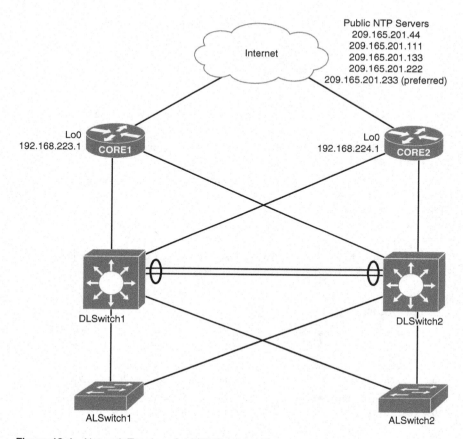

Figure 19-1 Network Topology for NTP Configuration

Core1 Router

`Core1(config)# ntp server 209.165.201.44`	Configures router to synchronize its clock to a public NTP server at address 209.165.201.44
`Core1(config)# ntp server 209.165.201.111`	Configures router to synchronize its clock to a public NTP server at address 209.165.201.111
`Core1(config)# ntp server 209.165.201.133`	Configures router to synchronize its clock to a public NTP server at address 209.165.201.133
`Core1(config)# ntp server 209.165.201.222`	Configures router to synchronize its clock to a public NTP server at address 209.165.201.222
`Core1(config)# ntp server 209.165.201.233 prefer`	Configures router to synchronize its clock to a public NTP server at address 209.165.201.233. This is the preferred NTP server
`Core1(config)# ntp max-associations 200`	Configures the maximum number of NTP peer-and-client associations that the router will serve

Core1(config)# clock timezone EDT -5	Sets time zone to eastern daylight time
Core1(config)# clock summer-time EDT recurring 2 Sun Mar 2:00 1 Sun Nov 2:00	Configures the system to automatically switch to summer time and to repeat on the same day
Core1(config)# ntp master 10	Configures the router to serve as a master clock if the external NTP server is not available
Core1(config)# access-list 1 permit 127.127.1.1	Sets access list to permit packets coming from 127.127.1.1
Core1(config)# access-list 2 permit 192.168.0.0 0.0.255.255	Sets access list to permit packets coming from 192.168.x.x
Core1(config)# ntp access-group peer 1	Configures Core1 to peer with any devices identified in access list 1
Core1(config)# ntp access-group serve-only 2	Configures Core1 to receive only time requests from devices specified in the access list

Core2 Router

Core2(config)# ntp server 209.165.201.44	Configures router to synchronize its clock to a public NTP server at address 209.165.201.44
Core2(config)# ntp server 209.165.201.111	Configures router to synchronize its clock to a public NTP server at address 209.165.201.111
Core2(config)# ntp server 209.165.201.133	Configures router to synchronize its clock to a public NTP server at address 209.165.201.133
Core2(config)# ntp server 209.165.201.222	Configures router to synchronize its clock to a public NTP server at address 209.165.201.222
Core2(config)# ntp server 209.165.201.233 prefer	Configures router to synchronize its clock to a public NTP server at address 209.165.201.233. This is the preferred NTP server
Core2(config)# ntp max-associations 200	Configures the maximum number of NTP peer-and-client associations that the router will serve
Core2(config)# clock timezone EDT -5	Sets time zone to eastern daylight time
Core2(config)# clock summer-time EDT recurring 2 Sun Mar 2:00 1 Sun Nov 2:00	Configures the system to automatically switch to summer time and to repeat on the same day
Core2(config)# ntp master 10	Configures the router to serve as a master clock if the external NTP server is not available

`Core2(config)# access-list 1 permit 127.127.1.1`	Sets access list to permit packets coming from 127.127.1.1
`Core2(config)# access-list 2 permit 192.168.0.0 0.0.255.255`	Sets access list to permit packets coming from 192.168.*x.x*
`Core2(config)# ntp access-group peer 1`	Configures Core2 to peer with any devices identified in access list 1
`Core2(config)# ntp access-group serve-only 2`	Configures Core2 to receive only time requests from devices specified in the access list

DLSwitch1

`DLSwitch1(config)# ntp server 192.168.223.1`	Configures DLSwitch1 to synchronize its clock to an NTP server at address 192.168.223.1
`DLSwitch1(config)# ntp server 192.168.224.1`	Configures DLSwitch1 to synchronize its clock to an NTP server at address 192.168.224.1
`DLSwitch1(config)# clock timezone EDT -5`	Sets time zone to eastern daylight time
`DLSwitch1(config)# clock summer-time EDT recurring 2 Sun Mar 2:00 1 Sun Nov 2:00`	Configures the system to automatically switch to summer time and to repeat on the same day

DLSwitch2

`DLSwitch2(config)# ntp server 192.168.223.1`	Configures DLSwitch2 to synchronize its clock to an NTP server at address 192.168.223.1
`DLSwitch2(config)# ntp server 192.168.224.1`	Configures DLSwitch2 to synchronize its clock to an NTP server at address 192.168.224.1
`DLSwitch2(config)# clock timezone EDT -5`	Sets time zone to eastern daylight time
`DLSwitch2(config)# clock summer-time EDT recurring 2 Sun Mar 2:00 1 Sun Nov 2:00`	Configures the system to automatically switch to summer time and to repeat on the same day

ALSwitch1

ALSwitch1(config)# **ntp server** **192.168.223.1**	Configures ALSwitch1 to synchronize its clock to an NTP server at address 192.168.223.1
ALSwitch1(config)# **ntp server 192.168.224.1**	Configures ALSwitch1 to synchronize its clock to an NTP server at address 192.168.224.1
ALSwitch1(config)# **clock timezone EDT -5**	Sets time zone to eastern daylight time
ALSwitch1(config)# **clock summer-time EDT recurring 2 Sun Mar 2:00 1 Sun Nov 2:00**	Configures the system to automatically switch to summer time and to repeat on the same day

ALSwitch2

ALSwitch2(config)# **ntp server 192.168.223.1**	Configures ALSwitch2 to synchronize its clock to an NTP server at address 192.168.223.1
ALSwitch2(config)# **ntp server 192.168.224.1**	Configures ALSwitch2 to synchronize its clock to an NTP server at address 192.168.224.1
ALSwitch2(config)# **clock timezone EDT -5**	Sets time zone to eastern daylight time
ALSwitch2(config)# **clock summer-time EDT recurring 2 Sun Mar 2:00 1 Sun Nov 2:00**	Configures the system to automatically switch to summer time and to repeat on the same day

CHAPTER 20

Layer Two Security Features

This chapter provides information and commands concerning the following topics:

- Setting passwords on a switch
- Configuring static MAC addresses
- Configuring switch port security
- Configuring sticky MAC addresses
- Verifying switch port security
- Recovering automatically from error-disabled ports
- Verifying autorecovery of error-disabled ports
- Configuring DHCP snooping
- Verifying DHCP snooping
- Configuring Dynamic ARP Inspection (DAI)
- Verifying Dynamic ARP Inspection
- Configuration example: Switch Security

Setting Passwords on a Switch

Setting passwords for switches is the same method as used for a router.

Switch2960(config)# **enable password cisco**	Sets the enable password to *cisco*
Switch2960(config)# **enable secret class**	Sets the encrypted secret password to *class*
Switch2960(config)# **line console 0**	Enters line console mode
Switch2960(config-line)# **login**	Enables password checking
Switch2960(config-line)# **password cisco**	Sets the password to *cisco*
Switch2960(config-line)# **exit**	Exits line console mode
Switch2960(config)# **line vty 0 15**	Enters line vty mode for all 16 virtual ports
Switch2960(config-line)# **login**	Enables password checking
Switch2960(config-line)# **password cisco**	Sets the password to *cisco*
Switch2960(config-line)# **exit**	Exits line vty mode
Switch2960(config)#	

Configuring Static MAC Addresses

Normally, switches learn of MAC addresses dynamically through the inspection of the source MAC address of incoming frames. These addresses then are placed in the Content Addressable Memory (CAM) table for future use. However, if required, you can manually add a MAC address to the CAM table—these are known as static MAC addresses. Static MAC addresses always overrule dynamic entries.

So why add static MAC addresses to the table? One reason could be to defeat a hacker who is trying to spoof a dynamically learned MAC address to change entries in the CAM table.

`Switch2960(config)# mac address-table static aaaa.aaaa.aaaa vlan 1 interface fastethernet 0/1`	Sets a permanent address to port fastethernet 0/1 in VLAN 1
`Switch2960(config)# no mac address-table static aaaa.aaaa.aaaa vlan 1 interface fastethernet 0/1`	Removes the permanent address to port fastethernet 0/1 in VLAN 1

Configuring Switch Port Security

`Switch(config)# interface fastethernet 0/1`	Moves to interface configuration mode
`Switch(config-if)# switchport mode access`	Sets the interface to access mode (as opposed to trunk mode) **NOTE:** A port cannot be in the default Dynamic Trunking Protocol (DTP) dynamic mode for port security to be enabled. It must be in either access or trunk mode
`Switch(config-if)# switchport port-security`	Enables port security on the interface
`Switch(config-if)# switchport port-security maximum 4`	Sets a maximum limit of four MAC addresses that will be allowed on this port **NOTE:** The maximum number of secure MAC addresses that you can configure on a switch is set by the maximum number of available MAC addresses allowed in the system
`Switch(config-if)# switchport port-security mac-address 1234.5678.90ab`	Sets a specific secure MAC address 1234.5678.90ab. You can add additional secure MAC addresses up to the maximum value configured
`Switch(config-if)# switchport port-security violation shutdown`	Configures port security to shut down the interface if a security violation occurs **NOTE:** In shutdown mode, the port is error-disabled, a log entry is made, and manual intervention or err-disable recovery must be used to reenable the interface

`Switch(config-if)# switch-port port-security violation restrict`	Configures port security to restrict mode if a security violation occurs **NOTE:** In restrict mode, frames from a nonallowed address are dropped, and a log entry is made. The interface remains operational
`Switch(config-if)# switchport port-security violation protect`	Configures port security to protect mode if a security violation occurs **NOTE:** In protect mode, frames from a nonallowed address are dropped, but no log entry is made. The interface remains operational

Configuring Sticky MAC Addresses

Sticky MAC addresses are a feature of port security. Sticky MAC addresses limit switch port access to a specific MAC address that can be dynamically learned, as opposed to a network administrator manually associating a MAC address with a specific switch port. These addresses are stored in the running configuration file. If this file is saved, the sticky MAC addresses do not have to be relearned when the switch is rebooted and thus provide a high level of switch port security.

`Switch(config)# interface fastethernet 0/5`	Moves to interface configuration mode
`Switch(config-if)# switchport port-security mac-address sticky`	Converts all dynamic port security learned MAC addresses to sticky secure MAC addresses
`Switch(config-if)# switchport port-security mac-address sticky vlan 10 voice`	Converts all dynamic port security learned MAC addresses to sticky secure MAC addresses on voice VLAN 10 **NOTE:** The **voice** keyword is available only if a voice VLAN is first configured on a port and if that port is not the access VLAN

Verifying Switch Port Security

`Switch# show port-security`	Displays security information for all interfaces
`Switch# show port-security interface fastethernet 0/5`	Displays security information for interface fastethernet 0/5
`Switch# show port-security address`	Displays all secure MAC addresses configured on all switch interfaces
`Switch# show mac address-table [dynamic]`	Displays the entire MAC address table Using the optional argument of **dynamic** will show only the dynamic addresses learned

Switch# **clear mac address-table dynamic**	Deletes all dynamic MAC addresses
Switch# **clear mac address-table dynamic address** *aaaa.bbbb.cccc*	Deletes the specified dynamic MAC address
Switch# **clear mac address-table dynamic interface fastethernet 0/5**	Deletes all dynamic MAC addresses on interface fastethernet 0/5
Switch# **clear mac address-table dynamic vlan 10**	Deletes all dynamic MAC addresses on VLAN 10
Switch# **clear mac address-table notification**	Clears MAC notification global counters **NOTE:** Beginning with Cisco IOS Release 12.1(11)EA1, the **clear mac address-table** command (no hyphen in **mac address**) replaces the **clear mac-address-table** command (with the hyphen in **mac-address**)

Recovering Automatically from Error-Disabled Ports

You can also configure a switch to autorecover error-disabled ports after a specified amount of time. By default, the autorecover feature is disabled.

Switch(config)# **errdisable recovery cause psecure-violation**	Enables the timer to recover from a port security violation disabled state
Switch(config)# **errdisable recovery interval** *seconds*	Specifies the time to recover from the error-disabled state. The range is 30 to 86,400 seconds. The default is 300 seconds **TIP:** Disconnect the offending host; otherwise, the port remains disabled, and the violation counter is incremented

Verifying Autorecovery of Error-Disabled Ports

Switch# **show errdisable recovery**	Displays error-disabled recovery timer information associated with each possible reason the switch could error-disable a port
Switch# **show interfaces status err-disabled**	Displays interface status or a list of interfaces in error-disabled state
Switch# **clear errdisable interface** *interface-id* **vlan** *[vlan-list]*	Reenables all or specified VLANs that were error-disabled on an interface

Configuring DHCP Snooping

Dynamic Host Configuration Protocol (DHCP) snooping is a DHCP security feature that provides network security by filtering untrusted DHCP messages and by building and maintaining a DHCP snooping binding database, which is also referred to as a DHCP snooping binding table.

`Switch(config)#` **`ip dhcp`** **`snooping`**	Enables DHCP snooping globally **NOTE:** If you enable DHCP snooping on a switch, the following DHCP relay agent commands are not available until snooping is disabled: `Switch(config)#`**`ip dhcp relay`** **`information check`** `Switch(config)#`**`ip dhcp relay`** **`information policy`** {**`drop`** | **`keep`** | **`replace`**} `Switch(config)#`**`ip dhcp relay`** **`information trust-all`** `Switch(config-if)#`**`ip dhcp relay`** **`information trusted`** If you enter these commands with DHCP snooping enabled, the switch returns an error message
`Switch(config)#` **`ip dhcp`** **`snooping vlan 20`**	Enables DHCP snooping on VLAN 20
`Switch(config)#` **`ip dhcp`** **`snooping vlan 10-35`**	Enables DHCP snooping on VLANs 10 through 35
`Switch(config)#` **`ip dhcp`** **`snooping vlan 20 30`**	Enables DHCP snooping on VLANs 20 through 30
`Switch(config)#` **`ip dhcp`** **`snooping vlan 10,12,14`**	Enables DHCP snooping on VLANs 10, 12, and 14
`Switch(config)#` **`ip dhcp`** **`snooping vlan 10-12,14`**	Enables DHCP snooping on VLANs 10 through 12 and VLAN 14
`Switch(config)#` **`ip dhcp snooping`** **`information option`**	Enables DHCP option 82 insertion **NOTE:** DHCP address allocation is usually based on an IP address, either the gateway IP address or the incoming interface IP address. In some networks, you might need additional information to determine which IP address to allocate. By using the "relay agent information option" (option 82), the Cisco IOS relay agent can include additional information about itself when forwarding DHCP packets to a DHCP server. The relay agent will add the circuit identifier suboption and the remote ID suboption to the relay information option and forward this all to the DHCP server

`Switch(config)# interface fastethernet0/1`	Moves to interface configuration mode
`Switch(config-if)# switchport mode trunk`	Forces the switch port to be a trunk
`Switch(config-if)# switchport trunk encapsulation dot1q`	Creates an uplink trunk with 802.1Q encapsulation **NOTE:** On newer switches that do not support ISL encapsulation, this command is not required (and may return an error as there is no need to choose between ISL and dot1q
`Switch(config-if)# switchport trunk allowed vlan 10,20`	Selects VLANs that are allowed transport on the trunk
`Switch(config-if)# ip dhcp snooping trust`	Configures the interface as trusted **NOTE:** There must be at least one trusted interface when working with DHCP snooping. It is usually the port connected to the DHCP server or to uplink ports. By default, all ports are untrusted
`Switch(config-if)# ip dhcp snooping limit rate 75`	Configures the number of DHCP packets per second that an interface can receive **NOTE:** The range of packets that can be received per second is 1 to 4,294,967,294. The default is no rate configured **TIP:** Cisco recommends an untrusted rate limit of no more than 100 packets per second
`Switch(config-if)# exit`	Returns to global configuration mode
`Switch(config)# ip dhcp snooping verify mac-address`	Configures the switch to verify that the source MAC address in a DHCP packet that is received on an untrusted port matches the client hardware address in the packet

Verifying DHCP Snooping

`Switch# show ip dhcp snooping`	Displays the DHCP snooping configuration for a switch
`Switch# show ip dhcp snooping binding`	Displays only the dynamically configured bindings in the DHCP snooping binding database
`Switch# show ip source binding`	Displays the dynamically and statically configured bindings
`Switch# show running-config`	Displays the status of the insertion and removal of the DHCP option 82 field on all interfaces

Configuring Dynamic ARP Inspection (DAI)

Dynamic ARP Inspection determines the validity of an ARP packet. This feature prevents attacks on the switch by not relaying invalid ARP requests and responses to other ports in the same VLAN.

`Switch(config)# ip dhcp snooping`	Enables DHCP snooping, globally
`Switch(config)# ip dhcp snooping vlan 10-20`	Enables DHCP snooping on VLANs 10 to 20
`Switch(config)# ip arp inspection vlan 10-20`	Enables DAI on VLANs 10 to 20, inclusive
`Switch(config)#` `ip arp inspection validate src-mac`	Configures DAI to drop ARP packets when the source MAC address in the body of the ARP packet does not match the source MAC address specified in the Ethernet header. This check is performed on both ARP requests and responses
`Switch(config)#` `ip arp inspection validate dst-mac`	Configures DAI to drop ARP packets when the destination MAC address in the body of the ARP packet does not match the destination MAC address specified in the Ethernet header. This check is performed on both ARP requests and responses
`Switch(config)#` `ip arp inspection validate ip`	Configures DAI to drop ARP packets that have invalid and unexpected IP addresses in the ARP body, such as 0.0.0.0, 255.255.255.255, or all IP multicast addresses. Sender IP addresses are checked in all ARP requests and responses, and target IP addresses are checked only in ARP responses
`Switch(config)# interface fastethernet0/24`	Moves to interface configuration mode
`Switch(config-if)#` `ip dhcp snooping trust`	Configures the interface as trusted for DHCP snooping
`Switch(config-if)#` `ip arp inspection trust`	Configures the connection between switches as trusted for DAI **NOTE:** By default, all interfaces are untrusted

TIP: It is generally advisable to configure all access switch ports as untrusted and to configure all uplink ports that are connected to other switches as trusted.

Verifying Dynamic ARP Inspection

`Switch# show ip arp inspection interfaces`	Verifies the dynamic ARP configuration
`Switch# show ip arp inspection vlan 10`	Verifies the dynamic ARP configuration for VLAN 10
`Switch# show ip arp inspection statistics vlan 10`	Displays the dynamic ARP inspection statistics for VLAN 10

Configuration Example: Switch Security

Figure 20-1 shows the network topology for the secure configuration of a 2960 series switch using commands covered in this chapter. Commands from other chapters are used as well.

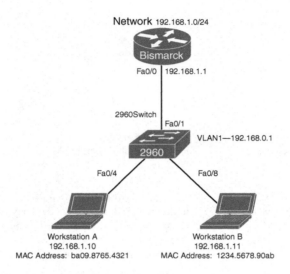

Figure 20-1   Network Topology for 2960 Series Switch Secure Configuration

`Switch>` **`enable`**	Enters privileged EXEC mode
`Switch#` **`configure terminal`**	Enters global configuration mode
`Switch(config)#` **`no ip domain-lookup`**	Turns off Domain Name System (DNS) queries so that spelling mistakes do not slow you down
`switch(config)#` **`hostname Switch2960`**	Sets the host name
`Switch2960(config)#` **`enable secret cisco`**	Sets the encrypted secret password to *cisco*
`Switch2960(config)#` **`line console 0`**	Enters line console mode
`Switch2960(config-line)#` **`logging synchronous`**	Appends commands to a new line; router information will not interrupt
`Switch2960(config-line)#` **`login`**	Requires user to log in to console before use
`Switch2960(config-line)#` **`password switch`**	Sets the password to *switch*
`Switch2960(config-line)#` **`exec-timeout 0 0`**	Prevents console from logging out due to lack of user input

`Switch2960(config-line)#` `exit`	Moves back to global configuration mode
`Switch2960(config)#` **`line`** **`vty 0 15`**	Moves to configure all 16 vty ports at the same time
`Switch2960(config-line)#` **`login`**	Requires user to log in to vty port before use
`Switch2960(config-line)#` **`password class`**	Sets the password to *class*
`Switch2960(config-line)#` **`exit`**	Moves back to global configuration mode
`Switch2960(config)#` **`ip`** **`default-gateway 192.168.1.1`**	Sets default gateway
`Switch2960(config)#` **`interface vlan 1`**	Moves to virtual interface VLAN 1 configuration mode
`Switch2960(config-if)#` **`ip address 192.168.1.2`** **`255.255.255.0`**	Sets the IP address and netmask for switch
`Switch2960(config-if)#` **`no`** **`shutdown`**	Turns the virtual interface on
`Switch2960(config-if)#` **`interface fastethernet 0/1`**	Moves to interface configuration mode for fastethernet 0/1
`Switch2960(config-if)#` **`description Link to`** **`Bismarck Router`**	Sets a local description
`Switch2960(config-if)#` **`interface fastethernet 0/4`**	Moves to interface configuration mode for fastethernet 0/4
`Switch2960(config-if)#` **`description Link to`** **`Workstation A`**	Sets a local description
`Switch2960(config-if)#` **`switchport mode access`**	Sets the interface to access mode
`Switch2960(config-if)#` **`switchport port-security`**	Activates port security
`Switch(config-if)#` **`switchport port-security`** **`mac-address sticky`**	Converts all dynamic port security learned MAC addresses to sticky secure MAC addresses
`Switch2960(config-if)#` **`switchport port-security`** **`maximum 1`**	Allows only one MAC address in the MAC table. This is the default number and not a required command, shown here for a visual reminder
`Switch2960(config-if)#` **`switchport port-security`** **`violation shutdown`**	Turns off port if more than one MAC address is reported. This is the default action and not a required command, shown here for a visual reminder

Switch2960(config-if)# **interface fastethernet 0/8**	Moves to interface configuration mode for fastethernet 0/8
Switch2960(config-if)# **description Link to Workstation B**	Sets a local description
Switch2960(config-if)# **switchport mode access**	Sets the interface to access mode
Switch2960(config-if)# **switchport port-security mac-address 1234.5678.90ab**	Sets a specific secure MAC address 1234.5678.90ab. You can add additional secure MAC addresses up to the maximum value configured
Switch2960(config-if)# **switchport port-security maximum 1**	Allows only one MAC address in the MAC table. This is the default number and not a required command, shown here for a visual reminder
Switch2960(config-if)# **switchport port-security violation shutdown**	Turns off port if more than one MAC address is reported. This is the default action and not a required command, shown here for a visual reminder
Switch2960(config-if)# **exit**	Returns to global configuration mode
Switch2960(config)# **exit**	Returns to privileged EXEC mode
Switch2960# **copy running-config startup-config**	Saves the configuration to NVRAM
Switch2960#	

CHAPTER 21

Managing Traffic Using Access Control Lists (ACLs)

This chapter provides information and commands concerning the following topics:

- Access list numbers
- Using wildcard masks
- ACL keywords
- Creating standard ACLs
- Applying standard ACLs to an interface
- Verifying ACLs
- Removing ACLs
- Creating extended ACLs
- Applying extended ACLs to an interface
- The **established** keyword
- The **log** keyword
- Creating named ACLs
- Using sequence numbers in named ACLs
- Removing specific lines in named ACLs using sequence numbers
- Sequence number tips
- Including comments about entries in ACLs
- Restricting virtual terminal access
- Tips for configuring ACLs
- IPv6 ACLs
- Verifying IPv6 ACLs
- Configuration examples: IPv4 ACLs
- Configuration examples: IPv6 ACLs

Access List Numbers

Although many different protocols can use access control lists (ACLs), the CCNA 200-301 certification exam is concerned only with IPv4 ACLs. The following chart shows some of the other protocols that can use ACLs.

1–99 or 1300–1999	Standard IPv4
100–199 or 2000–2699	Extended IPv4

NOTE: IPv6 ACLs do not use numbers; IPv6 ACLs are configured using names only.

Using Wildcard Masks

When applied to an IP address, a wildcard mask identifies which addresses get matched to be applied to the **permit** or **deny** argument in an ACL statement. A wildcard mask can identify a single host, a range of hosts, a complete network or subnetwork, or even all possible addresses.

There are two rules when working with wildcard masks:

- A 0 (zero) in a wildcard mask means to check the corresponding bit in the address for an exact match.

- A 1 (one) in a wildcard mask means to ignore the corresponding bit in the address—can be either 1 or 0. In the examples, this is shown as x.

Example 1: 172.16.0.0 0.0.255.255

> 172.16.0.0 = 10101100.00010000.00000000.00000000
>
> 0.0.255.255 = 00000000.00000000.11111111.11111111
>
> > Result = 10101100.00010000.*xxxxxxxx.xxxxxxxx*

172.16.*x.x* (Anything between 172.16.0.0 and 172.16.255.255 matches the example statement)

> **TIP:** An octet of all 0s means that the octet has to match exactly to the address. An octet of all 1s means that the octet can be ignored.

Example 2: 172.16.8.0 0.0.7.255

> 172.16.8.0 = 10101100.00010000.00001000.00000000
>
> 0.0.7.255 = 00000000.00000000.00000111.11111111
>
> > Result = 10101100.00010000.00001*xxx.xxxxxxxx*

> 00001*xxx* = 00001**000** to 00001**111** = 8–15
>
> *xxxxxxxx* = 00000000 to 11111111 = 0–255

Anything between 172.16.8.0 and 172.16.15.255 matches the example statement

ACL Keywords

`any`	Used in place of 0.0.0.0 255.255.255.255, matches any address that it is compared against
`host`	Used in place of 0.0.0.0 in the wildcard mask, matches only one specific address

Creating Standard ACLs

> **NOTE:** Standard ACLs are the oldest type of ACL. They date back as early as Cisco IOS Release 8.3. Standard ACLs control traffic by comparing the source of the IP packets to the addresses configured in the ACL.

NOTE: Each line in an ACL is called an access control entry (ACE). Many ACEs grouped together form a single ACL.

`Router(config)#`
`access-list 10`
`permit 172.16.0.0`
`0.0.255.255`
	Read this line to say, "All packets with a source IP address of 172.16.*x.x* will be matched by the statement, and the packet will be exited from processing the rest of the ACL"
	
	access-list = ACL command
	
	10 = Arbitrary number between 1 and 99, or 1300 and 1999, designating this as a standard IP ACL
	
	permit = Packets that match this statement will be allowed to continue
	
	172.16.0.0 = Source IP address to be compared to
	
	0.0.255.255 = Wildcard mask

`Router(config)#`
`access-list 10 deny`
`host 172.17.0.1`
	Read this line to say, "All packets with a source IP address of 172.17.0.1 will be dropped and discarded"

`access-list`	ACL command
`10`	Number between 1 and 99, or 1300 and 1999, designating this as a standard IP ACL
`deny`	Packets that match this statement will be dropped and discarded
`host`	Keyword
`172.17.0.1`	Specific host address

`Router(config)#`
`access-list 10`
`permit any`
	Read this line to say, "All packets with any source IP address will be matched by the statement, and the packet will be exited from processing the rest of the ACL"

`access-list`	ACL command
`10`	Number between 1 and 99, or 1300 and 1999, designating this as a standard IP ACL
`permit`	Packets that match this statement will be allowed to continue
`any`	Keyword to mean all IP addresses

TIP: An implicit **deny** statement is assumed into every ACL. You cannot see it, but it states "deny everything not already matched by an ACE in the list." This is always the last line of any ACL. If you want to defeat this implicit **deny**, put a **permit any** statement in your standard ACLs or a **permit ip any any** in your extended ACLs as the last line.

Applying Standard ACLs to an Interface

`Router(config)#` **interface** **gigabitethernet 0/0**	Moves to interface configuration mode
`Router(config-if)#` **ip access-group 10 out**	Takes all ACEs that are defined as being part of group 10 and applies them in an outbound manner. Packets leaving the router through interface gigabitethernet 0/0 will be checked

TIP: Access lists can be applied in either an inbound direction (keyword **in**) or an outbound direction (keyword **out**). Best practice is to have ACLs applied in an outbound direction.

TIP: Not sure in which direction to apply an ACL? Look at the flow of packets. Do you want to filter packets as they are going *in* a router's interface from an external source? Use the keyword **in** for this ACL. Do you want to filter packets before they go *out* of the router's interface toward another device? Use the keyword **out** for this ACL.

TIP: Apply a standard ACL as close as possible to the destination network or device. You do not want packets with the same source IP address to be filtered out early and prevented from reaching a legitimate destination.

Verifying ACLs

Router# **show ip interface**	Displays any ACLs applied to that interface
Router# **show access-lists**	Displays the contents of all ACLs on the router
Router# **show access-list** *access-list-number*	Displays the contents of the ACL by the *number* specified
Router# **show access-list** *name*	Displays the contents of the ACL by the *name* specified
Router# **show run**	Displays all ACLs and interface assignments

Removing ACLs

Router(config)# **no access-list 10**	Removes *all* ACEs in ACL number 10

Creating Extended ACLs

NOTE: Extended ACLs were also introduced in Cisco IOS Release 8.3. Extended ACLs control traffic by comparing the source and destination of the IP packets to the addresses configured in the ACL. Extended ACLs can also filter packets using protocol/port numbers for a more granular filter.

Router(config)# **access-list 110 permit tcp 172.16.0.0 0.0.0.255 192.168.100.0 0.0.0.255 eq 80**	Read this line to say, "HTTP packets with a source IP address of 172.16.0.*x* will be matched by the statement, and the packet will be exited from processing the rest of the ACL"
	access-list = ACL command
	110 = Number between 100 and 199, or 2000 and 2699, designating this as an extended IP ACL
	permit = Packets that match this statement will be allowed to continue
	tcp = Protocol must be TCP
	172.16.0.0 = Source IP address to be compared to
	0.0.0.255 = Wildcard mask for the source IP address

	192.168.100.0 = Destination IP address to be compared to
	0.0.0.255 = Wildcard mask for the destination IP address
	eq = Operand; means "equal to"
	80 = Port 80, indicating HTTP traffic
`Router(config)#` `access-list 110 deny` `tcp any 192.168.100.7` `0.0.0.0 eq 23`	Read this line to say, "Telnet packets with any source IP address will be dropped if they are addressed to specific host 192.168.100.7"
	access-list = ACL command
	110 = Number between 100 and 199, or 2000 and 2699, designating this as an extended IP ACL
	deny = Packets that match this statement will be dropped and discarded
	tcp = Protocol must be TCP protocol
	any = Any source IP address
	192.168.100.7 = Destination IP address to be compared to
	0.0.0.0 = Wildcard mask; address must match exactly
	eq = Operand, means "equal to"
	23 = Port 23, indicating Telnet traffic

Applying Extended ACLs to an Interface

`Router(config)#` **interface** **gigabitethernet 0/0**	Moves to interface configuration mode and takes all access list lines that are defined as being part of group 110 and applies them in an inbound manner. Packets going in gigabitethernet 0/0 will be checked
`Router(config-if)#` **ip** **access-group 110 in**	

TIP: Access lists can be applied in either an inbound direction (keyword **in**) or an outbound direction (keyword **out**). Best practice for extended ACLs is to apply them in an inbound manner.

TIP: Only one access list can be applied per interface, per direction, per protocol.

TIP: Apply an extended ACL as close as possible to the source network or device. This ensures that packets that are intended to be dropped are not allowed to travel.

The established Keyword

The **established** keyword is an optional keyword that is used with the TCP protocol only. It indicates an established connection. A match occurs only if the TCP segment has the ACK or RST control bits set.

`Router(config)#` **access-list 110 permit tcp** **172.16.0.0 0.0.0.255 eq 80 192.168.100.0** **0.0.0.255 established**	Indicates an established connection

TIP: The **established** keyword works only for TCP, not User Datagram Protocol (UDP).

TIP: Consider the following situation: You do not want hackers exploiting destination port 80 to access your network. Because you do not host a local web server (destination port 80), it is possible to block incoming (to your network) traffic on destination port 80, except that your internal users need web access. When they request a web page from the Internet, return traffic inbound on source port 80 must be allowed. The solution to this problem is to use the **established** command. The ACL allows the response to enter your network because it has the ACK bit set as a result of the initial request from inside your network. Requests from the outside world are blocked because the ACK bit is not set, but responses are allowed through.

The log Keyword

The **log** keyword is an optional parameter that causes an informational logging message about the packet matching the entry to be sent to the console. The log message includes the access list number, whether the packet was permitted or denied, the source address, the number of packets, and if appropriate, the user-defined cookie or router-generated hash value. The message is generated for the first packet that matches and then at 5-minute intervals, including the number of packets permitted or denied in the prior 5-minute interval.

CAUTION: ACL logging can be CPU intensive and can negatively affect other functions of the network device.

`Router(config)# access-list 1`
`permit 172.16.10.0 0.0.0.255`
`log`
	Indicates that logging will be enabled on this ACE

`Router(config)# access-list 1`
`permit 172.16.10.0 0.0.0.255`
`log SampleUserValue`
	Indicates that logging will be enabled on this ACE. The word *SampleUserValue* will be appended to each syslog entry

`Router(config)# access-list`
`110 permit tcp 172.16.0.0`
`0.0.0.255 192.168.100.0`
`0.0.0.255 eq 80 log`
	Indicates that logging will be enabled on this ACE

`Router(config)# access-list`
`110 permit tcp 172.16.0.0`
`0.0.0.255 192.168.100.0`
`0.0.0.255 eq 80 log-input`
	Indicates that logging will be enabled on this input and will include the input interface and source MAC address or virtual circuit in the logging output

`Router(config)# access-list`
`110 permit tcp 172.16.0.0`
`0.0.0.255 192.168.100.0`
`0.0.0.255 eq 80 log-input`
`SampleUserValue`
	Indicates that logging will be enabled on this ACE and will include the input interface and source MAC address or virtual circuit in the logging output. The word *SampleUserValue* will be appended to each syslog entry

TIP: The level of messages logged to the console is controlled by the **logging console** command.

TIP: After you specify the **log** keyword (and the associated word argument) or the **log-input** keyword (and the associated word argument), you cannot specify any other keywords or settings for this command.

TIP: The **log-input** keyword (and the associated word argument) is only available in extended ACLs for IPv4 or IPv6 ACLs.

Creating Named ACLs

Router(config)# **ip access-list extended serveraccess**	Creates an extended named ACL called *serveraccess* and moves to named ACL configuration mode
Router(config-ext-nacl)# **permit tcp any host 131.108.101.99 eq smtp**	Permits mail packets from any source to reach host 131.108.101.99
Router(config-ext-nacl)# **permit udp any host 131.108.101.99 eq domain**	Permits Domain Name System (DNS) packets from any source to reach host 131.108.101.99
Router(config-ext-nacl)# **deny ip any any log**	Denies all other packets from going anywhere. If any packets do get denied, this logs the results for you to look at later
Router(config-ext-nacl)# **exit**	Returns to global configuration mode
Router(config)# **interface gigabitethernet 0/0** Router(config-if)# **ip access-group serveraccess out**	Moves to interface configuration mode and applies this ACL to the gigabitethernet interface 0/0 in an outbound direction
Router(config)# **ip access-list standard teststandardacl**	Creates a standard-named ACL called *teststandardacl* and moves to named ACL configuration mode
Router(config-std-nacl)# **permit host 192.168.1.11**	Permits packets from source address 192.168.1.11
Router(config-std-nacl)# **exit**	Returns to global configuration mode
Router(config)# **interface gigabitethernet 0/1** Router(config-if)# **ip access-group teststandardacl out**	Moves to interface configuration mode and applies this ACL to the gigabitethernet interface 0/1 in an outbound direction

TIP: The prompt of the device changes according to whether the named ACL is standard (config-std-nacl) or extended (config-ext-nacl).

Using Sequence Numbers in Named ACLs

Router(config)# **ip access-list extended serveraccess2**	Creates an extended-named ACL called *serveraccess2*
Router(config-ext-nacl)# **10 permit tcp any host 131.108.101.99 eq smtp**	Uses sequence number 10 for this line
Router(config-ext-nacl)# **20 permit udp any host 131.108.101.99 eq domain**	Sequence number 20 will be applied after line 10

Router(config-ext-nacl)# **30 deny ip any any log**	Sequence number 30 will be applied after line 20
Router(config-ext-nacl)# **exit**	Returns to global configuration mode
Router(config)# **interface gigabitethernet 0/0**	Moves to interface configuration mode
Router(config-if)# **ip access-group serveraccess2 out**	Applies this ACL in an outbound direction
Router(config-if)# **exit**	Returns to global configuration mode
Router(config)# **ip access-list extended serveraccess2**	Moves to named ACL configuration mode for the ACL *serveraccess2*
Router(config-ext-nacl)# **25 permit tcp any host 131.108.101.99 eq ftp**	Sequence number 25 places this line after line 20 and before line 30
Router(config-ext-nacl)# **exit**	Returns to global configuration mode

TIP: Sequence numbers are used to allow for easier editing of your ACLs. The preceding example used numbers 10, 20, and 30 in the ACL lines. If you had needed to add another line to this ACL, it would have previously been added after the last line—line 30. If you had needed a line to go closer to the top, you would have had to remove the entire ACL and then reapply it with the lines in the correct order. Now you can enter a new line with a sequence number (this example used number 25), placing it in the correct location.

NOTE: The *sequence-number* argument was added in Cisco IOS Release 12.2(14)S. It was integrated into Cisco IOS Release 12.2(15)T.

Removing Specific Lines in Named ACLs Using Sequence Numbers

Router(config)# **ip access-list extended serveraccess2**	Moves to named ACL configuration mode for the ACL *serveraccess2*
Router(config-ext-nacl)# **no 20**	Removes line 20 from the list
Router(config-ext-nacl)# **exit**	Returns to global configuration mode

Sequence Number Tips

- Sequence numbers start at 10 and increment by 10 for each line.
- The maximum sequence number is 2147483647.
  - If you have an ACL that is so complex that it needs a number this big, I'd ask your boss for a raise.
- If you forget to add a sequence number, the line is added to the end of the list and assigned a number that is 10 greater than the last sequence number.

- If you enter an entry that matches an existing entry (except for the sequence number), no changes are made.

- If the user enters a sequence number that is already present, an error message of "Duplicate sequence number" displays. You have to reenter the line with a new sequence number.

- Sequence numbers are changed on a router reload to reflect the increment by 10 policy (see first tip in this section). If your ACL has numbers 10, 20, 30, 32, 40, 50, and 60 in it, on reload these numbers become 10, 20, 30, 40, 50, 60, 70.

- If you want to change the numbering sequence of your ACLs to something other than incrementing by 10, use the global configuration command **ip access-list resequence** *name/number start# increment#*:

  ```
  Router(config)# ip access-list resequence serveracces 1 2
  ```

  - This resets the ACL named *serveraccess* to start at 1 and increment by steps of 2 (1, 3, 5, 7, 9, and so on). The range for using this command is 1 to 2147483647.

- Sequence numbers cannot be seen when using the Router# **show running-config** or Router# **show startup-config** command. To see sequence numbers, use one of the following commands:

  ```
  Router# show access-lists
  Router# show access-lists list_name
  Router# show ip access-list
  Router# show ip access-list list_name
  ```

Including Comments About Entries in ACLs

Router(config)# **access-list 10 remark only Neo has access**	The remark command allows you to include a comment (limited to 100 characters)
Router(config)# **access-list 10 permit host 172.16.100.119**	Read this line to say, "Host 172.16.100.119 will be permitted through the internetwork"
Router(config)# **ip access-list extended telnetaccess**	Creates a named ACL called *telnetaccess* and moves to named ACL configuration mode
Router(config-ext-nacl)# **remark do not let Agent Smith have telnet**	The remark command allows you to include a comment (limited to 100 characters)
Router(config-ext-nacl)# **deny tcp host 172.16.100.153 any eq telnet**	Read this line to say, "Deny this specific host Telnet access to anywhere in the internetwork"

TIP: You can use the **remark** command in any of the IP numbered standard, IP numbered extended, or named IP ACLs.

TIP: You can use the **remark** command either before or after a **permit** or **deny** statement. Therefore, be consistent in your placement to avoid confusion about which line the **remark** statement is referring to.

Restricting Virtual Terminal Access

`Router(config)# access-list 2` `permit host 172.16.10.2`	Permits host from source address of 172.16.10.2 to telnet/SSH into this router based on where this ACL is applied
`Router(config)# access-list 2` `permit 172.16.20.0 0.0.0.255`	Permits anyone from the 172.16.20.*x* address range to telnet/SSH into this router based on where this ACL is applied **NOTE:** The implicit deny statement restricts anyone else from being permitted to telnet/SSH
`Router(config)# line vty 0 4`	Moves to vty line configuration mode
`Router(config-line)#` `access-class 2 in`	Applies this ACL to all five vty virtual interfaces in an inbound direction

TIP: When restricting access through Telnet, use the **access-class** command rather than the **access-group** command, which is used when applying an ACL to a physical interface.

CAUTION: Do not apply an ACL intending to restrict Telnet traffic on a physical interface. If you apply to a physical interface, *all* packets are compared to the ACL before it can continue on its path to its destination. This scenario can lead to a large reduction in router performance.

Tips for Configuring ACLs

- Each statement in an ACL is known as an ACE.
- Conversely, ACEs are commonly called ACL statements.
- The type of ACL determines what is filtered.
- Standard ACLs filter only on source IP address.
- Extended ACLs filter on source IP address, destination IP address, protocol number, and port number.
- Use only one ACL per interface, per protocol (IPv4 or IPv6), per direction.
- Place your most specific statements at the top of the ACL. The most general statements should be at the bottom of the ACL.
- The last test in any ACL is the implicit **deny** statement. You cannot see it, but it is there.
- Every ACL must have at least one **permit** statement. Otherwise, you will deny everything.
- Place extended ACLs as close as possible to the source network or device when applying ACLs to an interface.
- Place standard ACLs as close as possible to the destination network or device when applying ACLs to an interface.
- You can use numbers when creating a named ACL. The name you choose is the number. For example, **ip access-list extended 150** creates an extended ACL named 150.

- An ACL can filter traffic going through a router, depending on how the ACL is applied. Think of yourself as standing in the middle of the router.

  - Are you filtering traffic that is coming into the router toward you? If so, make the ACL an inbound one using the keyword **in**.

  - Are you filtering traffic that is going away from you and the router and toward another device? If so, make the ACL an outbound one using the keyword **out**.

- Access lists that are applied to interfaces do not filter traffic that originates from that router.

- When restricting access through Telnet, use the **access-class** command rather than the **access-group** command, which is used when applying an ACL to a physical interface.

IPv6 ACLs

ACLs can also be created in IPv6. The syntax for creating an IPv6 ACL is limited to named ACLs.

`Router(config)#` **`ipv6 access-list`** **`v6example`**	Creates an IPv6 ACL called *v6example* and moves to IPv6 ACL configuration mode
`Router(config-ipv6-acl)#` **`permit`** **`tcp 2001:db8:300:201::/32 eq`** **`telnet any`**	Permits the specified IPv6 address to telnet to any destination
`Router(config-ipv6-acl)#` **`deny tcp`** **`host 2001:db8:1::1 any log-input`**	Denies a specific IPv6 host. Attempts will be logged
`Router(config-ipv6-acl)#` **`exit`**	Returns to global configuration mode
`Router(config)#` **`interface`** **`gigabitethernet 0/0`**	Moves to interface configuration mode
`Router(config-if)#` **`ipv6`** **`traffic-filter v6example out`**	Applies the IPv6 ACL named *v6example* to the interface in an outbound direction

TIP: You use the **traffic-filter** keyword rather than the **access-group** keyword when assigning IPv6 ACLs to an interface.

TIP: Wildcard masks are not used in IPv6 ACLs. Instead, the prefix length is used.

TIP: You still use the **access-class** keyword to assign an IPv6 ACL to virtual terminal (vty) lines for restricting Telnet/SSH access.

Verifying IPv6 ACLs

`R1#` **`show ipv6`** **`access-list`**	Displays the configured statements, their matches, and the sequence number of all access lists

Configuration Examples: IPv4 ACLs

Figure 21-1 illustrates the network topology for the configuration that follows, which shows five ACL examples using the commands covered in this chapter.

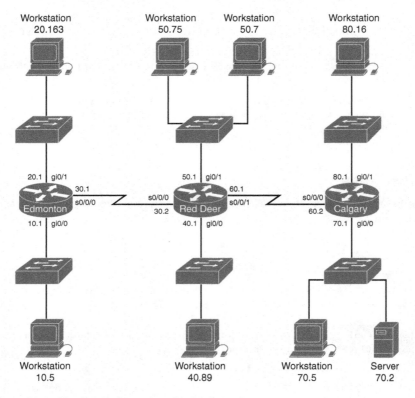

Figure 21-1   Network Topology for ACL Configuration

Example 1: Write an ACL that prevents the 10.0 network from accessing the 40.0 network but allows everyone else to access the 40.0 network.

RedDeer(config)# `access-list 10 deny 172.16.10.0 0.0.0.255`	The standard ACL denies the complete network for the complete TCP/IP suite of protocols
RedDeer(config)# `access-list 10 permit any`	Defeats the implicit deny
RedDeer(config)# `interface gigabitethernet 0/0`	Moves to interface configuration mode
RedDeer(config)# `ip access-group 10 out`	Applies ACL in an outbound direction

Example 2: Write an ACL that states that 10.5 cannot access 50.7. Everyone else can.

Edmonton(config)# **access list 115 deny ip host 172.16.10.5 host 172.16.50.7**	The extended ACL denies a specific host for the entire TCP/IP suite to a specific destination
Edmonton(config)# **access list 115 permit ip any any**	All others are permitted through
Edmonton(config)# **interface gigabitethernet 0/0**	Moves to interface configuration mode
Edmonton(config)# **ip access-group 115 in**	Applies the ACL in an inbound direction

Example 3: Write an ACL that states that 10.5 can telnet to the Red Deer router. No one else can.

RedDeer(config)# **access-list 20 permit host 172.16.10.5**	The standard ACL allows a specific host access. The implicit deny statement filters everyone else out
RedDeer(config)# **line vty 0 4**	Moves to vty configuration mode
RedDeer(config-line)# **access-class 20 in**	Applies ACL 20 in an inbound direction. Remember to use access-class, not access-group

Example 4: Write a named ACL that states that 20.163 can telnet to 70.2. No one else from 20.0 can telnet to 70.2. Any other host from any other subnet can connect to 70.2 using anything that is available.

Edmonton(config)# **ip access-list extended serveraccess**	Creates a named ACL and moves to named ACL configuration mode
Edmonton(config-ext-nacl)# **10 permit tcp host 172.16.20.163 host 172.16.70.2 eq telnet**	The specific host is permitted Telnet access to a specific destination
Edmonton(config-ext-nacl)# **20 deny tcp 172.16.20.0 0.0.0.255 host 172.16.70.2 eq telnet**	No other hosts are allowed to Telnet to the specified destination
Edmonton(config-ext-nacl)# **30 permit ip any any**	Defeats the implicit deny statement and allows all other traffic to pass through
Edmonton(config-ext-nacl)# **exit**	Returns to global configuration mode
Edmonton(config)# **interface gigabitethernet 0/1**	Moves to interface configuration mode
Edmonton(config-if)# **ip access-group serveraccess in**	Sets the ACL named *serveraccess* in an outbound direction on the interface

Example 5: Write an ACL that states that hosts 50.1 to 50.63 are not allowed web access to 80.16. Hosts 50.64 to 50.254 are. Everyone else can do everything else.

RedDeer(config)# **access-list 101 deny tcp 172.16.50.0 0.0.0.63 host 172.16.80.16 eq 80**	Creates an ACL that denies HTTP traffic from a range of hosts to a specific destination
RedDeer(config)# **access-list 101 permit ip any any**	Defeats the implicit deny statement and allows all other traffic to pass through
RedDeer(config)# **interface gigabitethernet 0/1**	Moves to interface configuration mode
RedDeer(config)# **ip access-group 101 in**	Applies the ACL in an inbound direction

Configuration Examples: IPv6 ACLs

Figure 21-2 shows the network topology for the configuration that follows, which demonstrates how to configure IPv6 ACLs. Assume that all basic configurations are accurate. The objective here is to create an ACL that acts as a firewall allowing HTTP, HTTPS, DNS, and Internet Control Message Protocol (ICMP) traffic to return from the Internet.

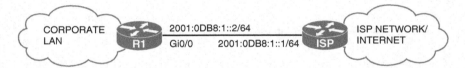

Figure 21-2   Configure IPv6 ACLs

R1(config)# **ipv6 access-list FIREWALL**	Creates a named extended IPv6 access list called FIREWALL and moves to IPv6 access list configuration mode
R1(config-ipv6-acl)# **permit tcp any eq www any established**	Permits HTTP traffic to return to the corporate LAN from the Internet if that traffic was originally sourced from the corporate LAN
R1(config-ipv6-acl)# **permit tcp any eq 443 any established**	Permits HTTPS traffic to return to the corporate LAN from the Internet if that traffic was originally sourced from the corporate LAN
R1(config-ipv6-acl)# **permit udp any eq domain any**	Permits DNS responses to return to the corporate LAN from the Internet
R1(config-ipv6-acl)# **permit icmp any any echo-reply**	Permits ICMP ping responses to return to the corporate LAN from the Internet

`R1(config-ipv6-acl)# permit icmp any any packet-too-big`	Permits ICMP Packet Too Big messages to return to the corporate LAN from the Internet **NOTE:** In IPv6, maximum transmission unit (MTU) discovery has moved from the router to the hosts. It is important to allow Packet Too Big messages to flow through the router to allow hosts to detect whether fragmentation is required
`R1(config-ipv6-acl)# exit`	Returns to global configuration mode
`R1(config)# interface gigabitethernet0/0`	Enters GigabitEthernet0/0 interface configuration mode
`R1(config-if)# ipv6 traffic-filter FIREWALL in`	Applies the IPv6 access list named FIREWALL to the interface in the inbound direction

NOTE: The "implicit deny" rule has changed for IPv6 access lists to take into account the importance of the Neighbor Discovery Protocol (NDP). NDP is to IPv6 what Address Resolution Protocol (ARP) is to IPv4, so naturally the protocol should not be disrupted. That is the reason two additional implicit statements have been added before the "implicit deny" statement at the end of each IPv6 ACL.

These implicit rules are as follows:

```
permit icmp any any nd-na
permit icmp any any nd-ns
deny ipv6 any any
```

It is important to understand that any explicit **deny ipv6 any any** statement overrides all three implicit statements, which can lead to problems because NDP traffic is blocked.

CHAPTER 22

Device Monitoring and Hardening

This chapter provides information about the following topics:

- Device monitoring
- Configuration backups
- Implementing logging
    - Configuring syslog
    - Syslog message format
    - Syslog severity levels
    - Syslog message example
- Device hardening
    - Configuring passwords
    - Password encryption
    - Password encryption algorithm types
    - Configuring SSH
    - Verifying SSH
    - Restricting virtual terminal access
    - Disabling unused services

Device Monitoring

Network administrators need to be able to perform more than just the configuration of network devices. They need to be able to monitor network devices to ensure that the network is operating as efficiently as possible and to identify potential bottlenecks or trouble spots. The following sections deal with protocols that can help monitor a network.

Configuration Backups

It is important to keep a copy of a router's configuration in a location other than NVRAM. Automated jobs can be set up to copy configurations from the router at regular intervals to local or remote file systems.

`Edmonton(config)# archive`	Enters archive configuration mode

`Edmonton(config-archive)#`
`path ftp://admin:cisco123@`
`192.168.10.3/$h.cfg`
	Sets the base file path for the remote location of the archived configuration
	
	The FTP server is located at 192.168.10.3
	
	The username to access the FTP server is *admin*
	
	The password is *cisco123*
	
	The path can be a local or remote path
	
	Path options include **flash**, **ftp**, **http**, **https**, **rcp**, **scp**, or **tftp**
	
	Two variables can be used with the **path** command:
	
	$h will be replaced with the device host name
	
	$t will be replaced with the date and time of the archive
	
	If you do not use $t, the names of the new files will be appended with a version number to differentiate from the previous configurations from the same device

`Edmonton(config-archive)#` `time-period 1440`	Sets the period of time (in minutes) in which to automatically archive the running-config. This number can range from 1 to 525,600 minutes. 1440 minutes = 1 day. 525,600 minutes = 1 year
`Edmonton(config-archive)#` `write-memory`	Enables automatic backup generation during write memory
`Edmonton# show archive`	Displays the list of archives. This command also has a pointer to the most recent archive

TIP: To create an archive copy manually, use the **archive config** command from EXEC mode:

```
Edmonton# archive config
```

TIP: When the **write-memory** command is enabled, the **copy running-config startup-config** command triggers an archive to occur.

Implementing Logging

Network administrators should implement logging to get insight into what is occurring in their network. When a router reloads, all local logs are lost, so it is important to implement logging to an external destination. These next sections deal with the different mechanisms that you can use to configure logging to a remote location.

Configuring Syslog

Edmonton(config)# **logging on**	Enables logging to all supported destinations
Edmonton(config)# **logging 192.168.10.53**	Sends logging messages to a syslog server host at address 192.168.10.53
Edmonton(config)# **logging sysadmin**	Sends logging messages to a syslog server host named *sysadmin*
Edmonton(config)# **logging trap x**	Sets the syslog server logging level to value *x*, where *x* is a number between 0 and 7 or a word defining the level (Table 22-1 provides more detail)
Edmonton(config)# **service sequence-numbers**	Stamps syslog messages with a sequence number
Edmonton(config)# **service timestamps log datetime**	Includes a time stamp on syslog messages
Edmonton(config)# **service timestamps log datetime msec**	Includes a time stamp measured in milliseconds on syslog messages

Syslog Message Format

The general format of syslog messages generated on Cisco IOS Software is as follows:

```
seq no:timestamp: %facility-severity-MNEMONIC:description
```

Item in Syslog Message	Definition
seq no	Sequence number. Stamped only if the **service sequence-numbers** global configuration command is configured
timestamp	Date and time of the message. Appears only if the **service timestamps log datetime** global configuration command is configured
facility	The facility to which the message refers (SNMP, SYS, and so on)
severity	Single-digit code from 0 to 7 that defines the severity of the message (see Table 22-1 for descriptions of the levels)
MNEMONIC	String of text that uniquely defines the message
description	String of text that contains detailed information about the event being reported

Syslog Severity Levels

Table 22-1 shows the eight levels of severity in logging messages.

TABLE 22-1   Syslog Severity Levels

Level #	Level Name	Description
0	Emergencies	System unusable
1	Alerts	Immediate action needed
2	Critical	Critical conditions
3	Errors	Error conditions
4	Warnings	Warning conditions
5	Notifications	Normal but significant conditions
6	Informational	Informational messages (default level)
7	Debugging	Debugging messages

Setting a level means you will get that level and everything numerically below it. Level 6 means you will receive messages for levels 0 through 6.

Syslog Message Example

The easiest syslog message to use as an example is the one that shows up every time you exit from global configuration mode back to privileged EXEC mode. You have just finished entering a command, and you want to save your work, but after you type **exit** you see something like this:

```
Edmonton(config)# exit
Edmonton#
*Jun 23:22:45:20.878: %SYS-5-CONFIG_I: Configured from console by
  console
Edmonton#
```

(Your output will differ depending on whether you have sequence numbers or time/date stamps configured.)

So what does this all mean?

- No sequence number is part of this message.
- The message occurred at June 23, at 22:45:20.878 (or 10:45 PM, and 20.878 seconds).
- It is a sys message, and it is level 5 (a notification).
- It is a config message; specifically, the configuration occurred from the console.

Device Hardening

Device security is critical to network security. A compromised device can cause the network to be compromised on a larger scale. The following sections deal with different ways to secure your Cisco IOS devices.

Configuring Passwords

These commands work on both routers and switches.

`Edmonton(config)# enable password cisco`	Sets the enable password to *cisco*. This password is stored as clear text
`Edmonton(config)# enable secret class`	Sets the enable secret password to *class*. This password is stored using a cryptographic hash function (SHA-256)
`Edmonton(config)# line console 0`	Enters console line mode
`Edmonton(config-line)# password console`	Sets the console line mode password to *console*
`Edmonton(config-line)# login`	Enables password checking at login
`Edmonton(config)# line vty 0 4`	Enters the vty line mode for all five vty lines
`Edmonton(config-line)# password telnet`	Sets the vty password to *telnet*
`Edmonton(config-line)# login`	Enables password checking at login
`Edmonton(config)# line aux 0`	Enters auxiliary line mode
`Edmonton(config-line)# password backdoor`	Sets auxiliary line mode password to *backdoor*
`Edmonton(config-line)# login`	Enables password checking at login

CAUTION: The **enable** *password* is not encrypted; it is stored as clear text. For this reason, recommended practice is that you *never* use the **enable** *password* command. Use only the **enable secret** *password* command in a router or switch configuration.

TIP: You can set both **enable secret** *password* and **enable** *password* to the same password. However, doing so defeats the use of encryption.

CAUTION: Line passwords are stored as clear text. They should be encrypted using the **service password-encryption** command at a bare minimum. However, this encryption method is weak and easily reversible. It is therefore recommended to enable authentication by the **username** command with the **secret** option because the **password** option to the **username** command still stores information using clear text.

TIP: The best place to store passwords is on an external authentication, authorization, and accounting (AAA) server.

Password Encryption

Edmonton(config)# **service password-encryption**	Applies a Vigenere cipher (type 7) weak encryption to passwords
Edmonton(config)# **enable password cisco**	Sets the enable password to *cisco*
Edmonton(config)# **line console 0**	Moves to console line mode
Edmonton(config-line)# **password cisco**	Continue setting passwords as above
	...
Edmonton(config)# **no service password-encryption**	Turns off password encryption

CAUTION: If you have turned on service password encryption, used it, and then turned it off, any passwords that you have encrypted stay encrypted. New passwords remain unencrypted.

TIP: The **service password-encryption** command works on the following passwords:

- Username
- Authentication key
- Privileged command
- Console
- Virtual terminal line access
- BGP neighbors

Passwords using this encryption are shown as type 7 passwords in the router configuration:

```
Edmonton# show running-config
                 <output omitted>
enable secret 4 Rv4kArhts7yA2xd8BD2YTVbts (4 signifies SHA-256 hash)
<output omitted>
line con 0
  password 7 00271A5307542A02D22842 (7 signifies Vigenere cipher)
line vty 0 4
  password 7 00271A5307542A02D22842 (7 signifies Vigenere cipher)
<output omitted>
R1#
```

Password Encryption Algorithm Types

There are different algorithm types available to hash a password in Cisco IOS:

- Type 4: Specified an SHA-256 encrypted secret string
  - Deprecated due to a software bug that allowed this password to be viewed in plaintext under certain conditions
- Type 5: Specifies a message digest algorithm 5 (MD5) encrypted secret

- Type 8: Specifies a Password-Based Key Derivation Function 2 with SHA-256 hashed secret (PBKDF2 with SHA-256)

- Type 9: Specifies a scrypt hashed secret (SCRYPT)

TIP: Configure all secret passwords using type 8 or type 9.

Edmonton(config)# **username demo8 algorithm-type shaw256 secret cisco**	Generates password encrypted with a type 8 algorithm
Edmonton(config)# **username demo9 algorithm-type scrypt secret cisco**	Generates password encrypted with a type 9 algorithm

NOTE: Type 5, 8, and 9 passwords are not reversible.

CAUTION: If you configure type 8 or type 9 passwords and then downgrade to an IOS release that does not support type 8 and type 9 passwords, you must configure the type 5 passwords before downgrading. If not, you are locked out of the device and a password recovery is required.

Configuring SSH

Although Telnet is the default way of accessing a router, it is the most unsecure way. Secure Shell (SSH) provides an encrypted alternative for accessing a router.

CAUTION: SSH Version 1 implementations have known security issues. It is recommended to use SSH Version 2 whenever possible.

NOTE: The device name cannot be the default *Switch* (on a switch) or *Router* (on a router). Use the **hostname** command to configure a new host name of the device.

NOTE: The Cisco implementation of SSH requires Cisco IOS Software to support Rivest, Shamir, Adleman (RSA) authentication and minimum Data Encryption Standard (DES) encryption (a cryptographic software image).

Edmonton(config)# **username Roland password tower**	Creates a locally significant username/password combination. These are the credentials you must enter when connecting to the router with SSH client software
Edmonton(config)# **username Roland privilege 15 secret tower**	Creates a locally significant username of *Roland* with privilege level 15. Assigns a secret password of *tower*
Edmonton(config)# **ip domain-name test.lab**	Creates a host domain for the router
Edmonton(config)# **crypto key generate rsa modulus 2048**	Enables the SSH server for local and remote authentication on the router and generates an RSA key pair. The number of modulus bits on the command line is 2048. The size of the key modulus is 360 to 4096 bits

`Edmonton(config)#` **`ip ssh`** **`version 2`**	Enables SSH version 2 on the device **NOTE:** To work, SSH requires a local username database, a local IP domain, and an RSA key to be generated
`Edmonton(config)#` **`line vty 0 4`**	Moves to vty configuration mode for all five vty lines of the router **NOTE:** Depending on the IOS release and platform, there may be more than five vty lines
`Edmonton(config-line)#` **`login local`**	Enables password checking on a per-user basis. The username and password will be checked against the data entered with the username global configuration command
`Edmonton(config-line)#` **`transport input ssh`**	Limits remote connectivity to SSH connections only—disables Telnet

Verifying SSH

`Edmonton#` **`show ip ssh`**	Verifies that SSH is enabled
`Edmonton#` **`show ssh`**	Checks the SSH connection to the device

Restricting Virtual Terminal Access

`Edmonton(config)#` **`access-list 2 permit`** **`host 172.16.10.2`**	Permits host from source address of 172.16.10.2 to telnet/SSH into this router based on where this ACL is applied
`Edmonton(config)#` **`access-list 2 permit`** **`172.16.20.0 0.0.0.255`**	Permits anyone from the 172.16.20.x address range to telnet/SSH into this router based on where this ACL is applied **NOTE:** The implicit deny statement restricts anyone else from being permitted to Telnet/SSH
`Edmonton(config)#` **`access-list 2 deny`** **`any log`**	Any packets that are denied by this ACL are logged for review at a later time. This line is used instead of the implicit deny line
`Edmonton(config)#` **`line vty 0 4`**	Moves to vty line configuration mode **NOTE:** Depending on the IOS release and platform, there may be more than five vty lines
`Edmonton(config-line)#` **`access-class 2 in`**	Applies this ACL to all vty virtual interfaces in an inbound direction

TIP: When restricting access on vty lines, use the **access-class** command rather than the **access-group** command, which is used when applying an ACL to a physical interface.

CAUTION: Do not apply an ACL intending to restrict vty traffic on a physical interface. If you apply it to a physical interface, *all* packets are compared to the ACL before it can continue on its path to its destination. This can lead to a large reduction in router performance. An ACL on a physical interface has to specify the SSH or Telnet port number that you are trying to deny, in addition to identifying all the router's addresses that you could potentially SSH/Telnet to.

Disabling Unneeded Services

Services that are not being used on a router can represent a potential security risk. If you do not need a specific service, you should disable it.

TIP: If a service is off by default, disabling it does not appear in the running configuration.

TIP: Do not assume that a service is disabled by default; you should explicitly disable all unneeded services, even if you think they are already disabled.

TIP: Depending on the IOS Software release, some services are on by default; some are off. Be sure to check the IOS configuration guide for your specific software release to determine the default state of the service.

Table 22-2 lists the services that you should disable if you are not using them.

TABLE 22-2 Disabling Unneeded Services

Service	Command Used to Disable Service
DNS name resolution	Edmonton(config)# **no ip domain-lookup**
Cisco Discovery Protocol (CDP) (globally)	Edmonton(config)# **no cdp run**
CDP (on a specific interface)	Edmonton(config-if)# **no cdp enable**
Network Time Protocol (NTP)	Edmonton(config-if)# **ntp disable**
BOOTP server	Edmonton(config)# **no ip bootp server**
Dynamic Host Configuration Protocol (DHCP)	Edmonton(config)# **no service dhcp**
Proxy Address Resolution Protocol (ARP)	Edmonton(config-if)# **no ip proxy-arp**
IP source routing	Edmonton(config)# **no ip source-route**
IP redirects	Edmonton(config-if)# **no ip redirects**
HTTP service	Edmonton(config)# **no ip http server**

CHAPTER 23

Configuring and Securing a WLAN AP

This chapter provides information concerning the following topics:

- Initial Setup of a Wireless LAN Controller (WLC)
- Monitoring the WLC
- Configuring a VLAN (dynamic) interface
- Configuring a DHCP scope
- Configuring a WLAN
- Defining a RADIUS server
- Exploring management options
- Configuring a WLAN using WPA2 PSK

Initial Setup of a Wireless LAN Controller (WLC)

As an alternative to using the CLI to configure your WLC, you can use the simplified controller provisioning feature. This feature is enabled after first boot from a nonconfigured WLC, temporarily provides Dynamic Host Configuration Protocol (DHCP) service on the service port segment, and assigns PC clients a limited network address. The client can connect to the WLC using a web browser.

With a client PC connected to the WLC mapped service port, it gets an address from a limited range of 192.168.1.3 through 192.168.1.14. The WLC is assigned a fixed 192.168.1.1 address. Open a browser and connect to http://192.168.1.1 and the simplified setup wizard will help you to navigate through the minimal steps to fully configure the WLC. Figure 23-1 shows the initial page where you create the admin account, providing the admin username and password. Click **Start** to continue.

Figure 23-1   Simplified Setup Start Page

On the next page of the simplified setup wizard, shown in Figure 23-2, set up the WLC with a system name, country, date/time (automatically taken from the client PC clock), and NTP server. Also define the management IP address, subnet mask, default gateway, and VLAN ID for the management interface. Click **Next** to continue.

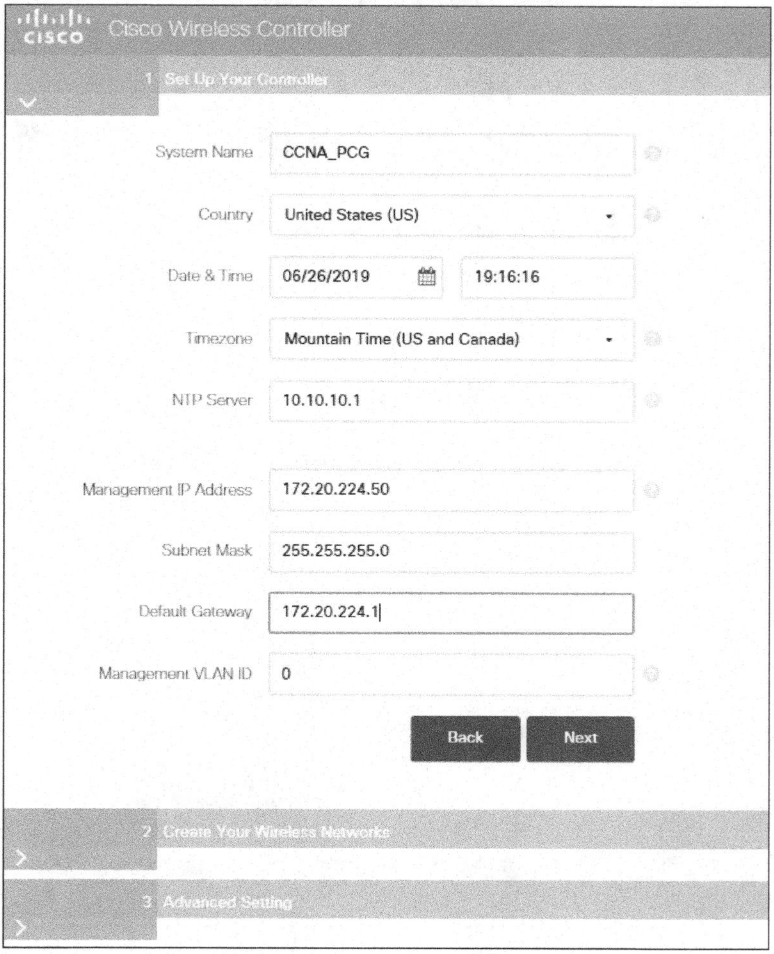

Figure 23-2   Set Up Your Controller Wizard Page

On the Create Your Wireless Networks page, shown in Figure 23-3, create the wireless network SSID, choose the security type, and specify the network/VLAN assignment. Optional is the inclusion of a Guest Network setup, a step you can use to add secure guess access with a separate network and access method for guests. Click **Next** to continue.

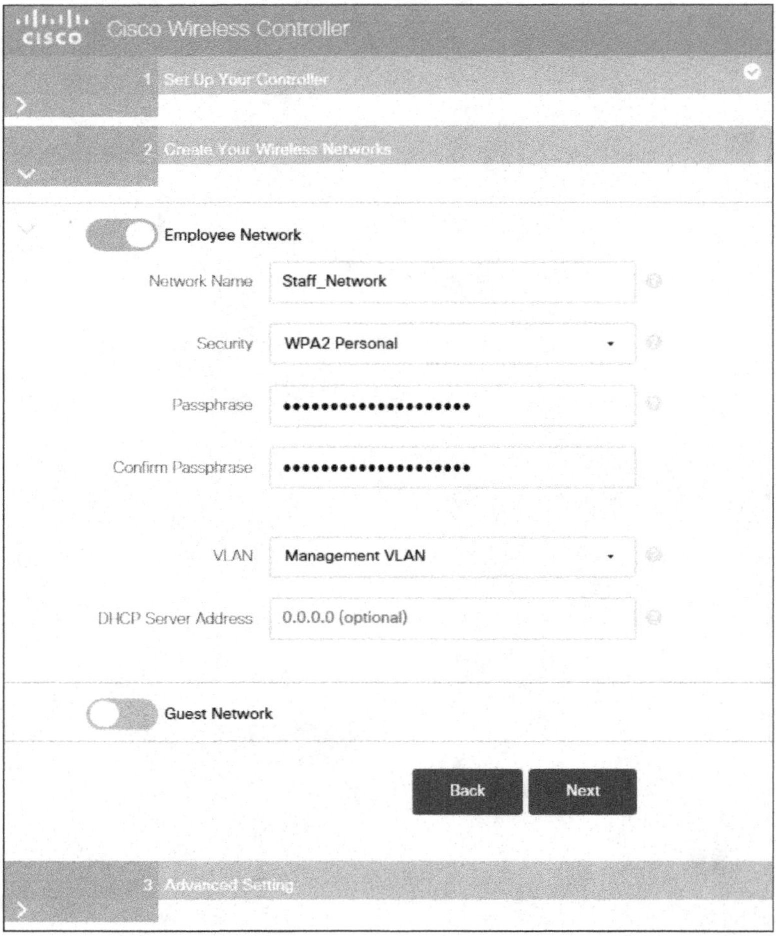

Figure 23-3   Create Your Wireless Networks Wizard Page

NOTE: From a security standpoint, it is advisable to configure WLANs with WPA2 with AES encryption, and 802.1X authentication. Avoid other security policies such as Open, WEP, WPA/TKIP, etc., unless absolutely needed for legacy client support. Using a pre-shared key as authentication is not recommended for enterprise environments, and should only be used for specific client compatibility scenarios. In these cases, a shared secret of 18 characters or more is advisable.

The next step of the wizard, shown in Figure 23-4, is to configure the WLC for intended RF use, and to take advantage of Cisco Wireless LAN Controller best practices defaults. Click **Next** to finalize the setup.

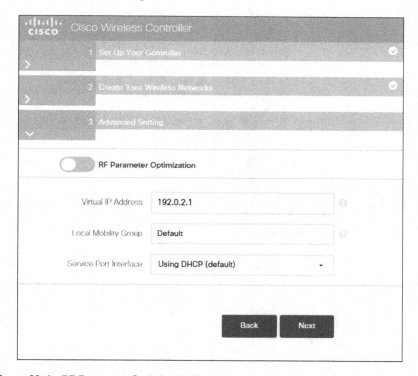

Figure 23-4   RF Parameter Optimization Settings

The simplified setup wizard will summarize the details of your configuration, as shown in Figure 23-5. Click **Apply** to save the configuration and reboot the WLC.

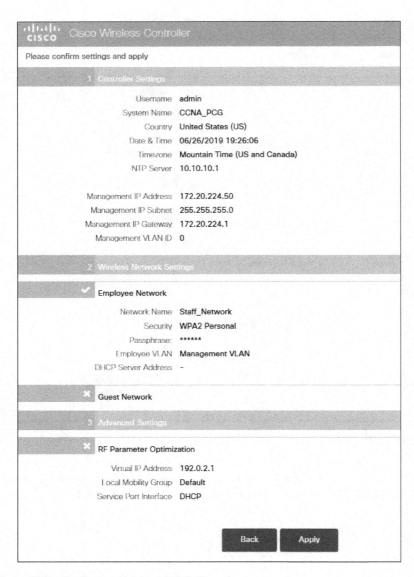

Figure 23-5   Confirming and Applying Settings

Figure 23-6 shows the reboot warning popup window. Click **OK**.

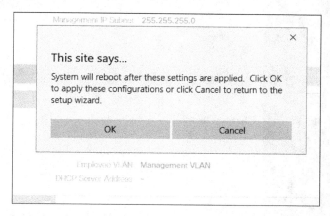

Figure 23-6   Rebooting the System

Once the WLC reboots, it disables the simplified set feature. You now need to use the management IP address of the WLC to log in. Figure 23-7 shows me logging in using the address of the service port.

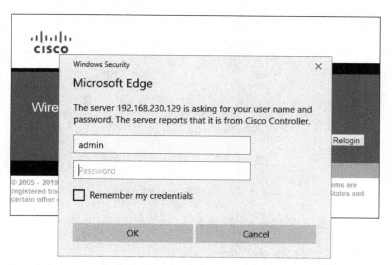

Figure 23-7   Logging in to the WLC

Monitoring the WLC

After logging in to your WLC, you see the Network Summary screen of the WLC, shown in Figure 23-8. You can't make any configuration changes from this screen; it is for viewing what the WLC is reporting. Click **Advanced** in the top-right corner.

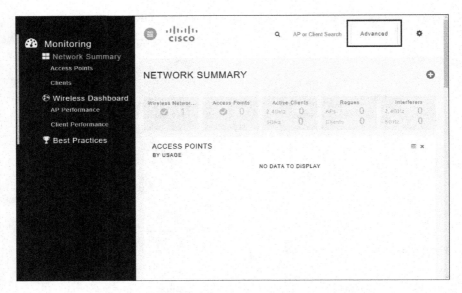

Figure 23-8   Navigating to the Advanced Monitor Summary Screen

Figure 23-9 shows the Advanced Monitor Summary screen, where you can drill down to pages to make changes to your configuration.

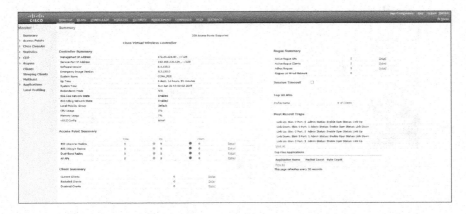

Figure 23-9   Advanced Monitor Summary Screen

Configuring a VLAN (Dynamic) Interface

Dynamic interfaces on a WLC are also known as VLAN interfaces. They are the same as VLANs for wireless clients. These interfaces allow for separate communication streams to exist on a controller's distribution system ports. Each dynamic interface is mapped to a wireless network, which is then mapped to a VLAN network.

Starting at the Advanced Monitor Summary screen, click **Controller** and then click **Interfaces**, as shown in Figure 23-10.

Figure 23-10   Controller Interfaces Screen

Click the **New** button, as shown in Figure 23-11, to create a new interface (scroll to the right side of the screen if you can't see the New button).

Figure 23-11   Creating a New Interface

From here you will step through the screens to enter the appropriate information for creating a new interface. Pay attention to capitalization and spacing. Figure 23-12 shows the first screen. Complete the fields and click **Apply**.

Figure 23-12   Naming a New Interface

Figure 23-13 shows the next screen for configuring this new interface. Under Physical Information, enter the port number of the interface. Remember that the physical RJ-45 connections are ports and that one or more interfaces can be configured on a port.

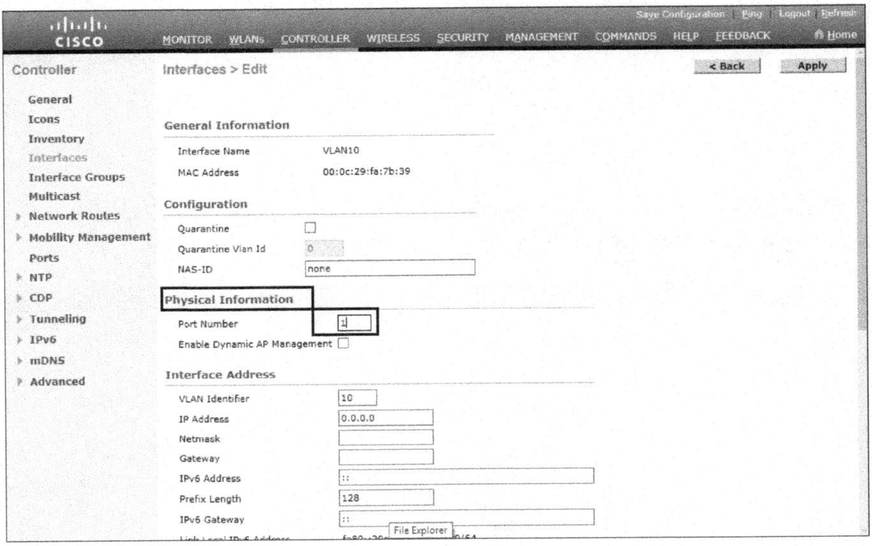

Figure 23-13   Entering the Port Number

Continue to the Interface Address section and enter the required IP address, netmask, and gateway information for the interface. In the DHCP Information section, enter the required DHCP server information. These sections are shown in Figure 23-14.

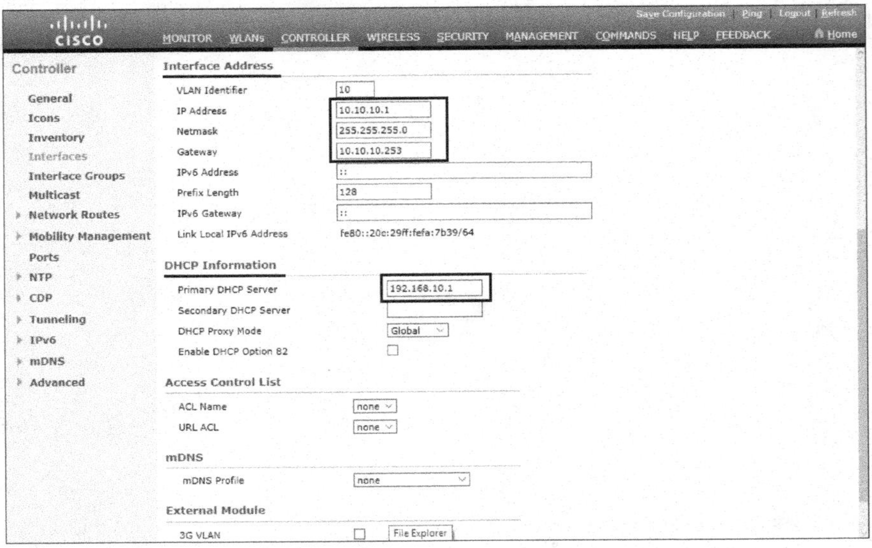

Figure 23-14   Interface Address and DHCP Information

After you have entered all the required information, scroll up and click **Apply** in the upper-right corner of the screen. A warning will appear reminding you that changing interface parameters may cause temporary loss of connectivity for some clients as the changes are being applied. Click **OK**, as shown in Figure 23-15.

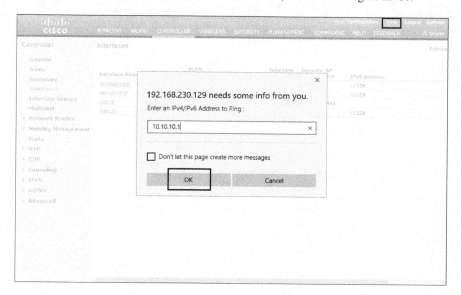

Figure 23-15   Applying Interface Changes

If you wish to test your configuration, click the **Ping** button in the upper-right part of the screen. Enter your new interface address and click **OK**, as shown in Figure 23-16.

Figure 23-16   Testing the New Interface

Figure 23-17 shows a successful ping to the interface address.

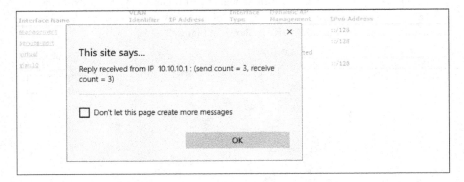

Figure 23-17   Ping Success Message

Configuring a DHCP Scope

DHCP will be used to provide IP addresses to the hosts that will be on the VLAN that you create. DHCP can be deployed from many different platforms. This example shows how to deploy DHCP from the WLC.

Starting on the Advanced Monitor Summary screen, click the **Controller** option in the top menu bar and then, in the navigation bar on the left side of the screen, expand the Internal DHCP Server option by clicking the blue triangle to its left, as shown in Figure 23-18.

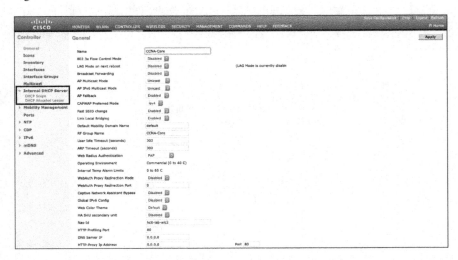

Figure 23-18   Internal DHCP Server Option

You have two choices under Internal DHCP Server: DHCP Scope and DHCP Allocated Leases. Click **DHCP Scope** and the corresponding page shows that you already have a configured DHCP scope, as shown in Figure 23-19: the day0-dhcp-mgmt scope that is used with the service port during initial configuration.

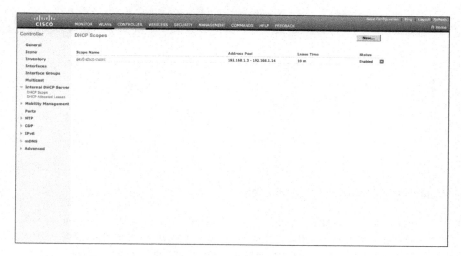

Figure 23-19 DHCP Scopes

To create a new DHCP scope, scroll (if necessary) to the right of the screen and click **New** in the upper-right corner, as shown in Figure 23-20.

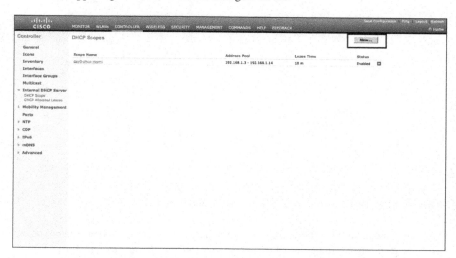

Figure 23-20 Selecting a New DHCP Scope

On the next screen, shown in Figure 23-21, enter the name of your new DHCP scope and then click the **Apply** button.

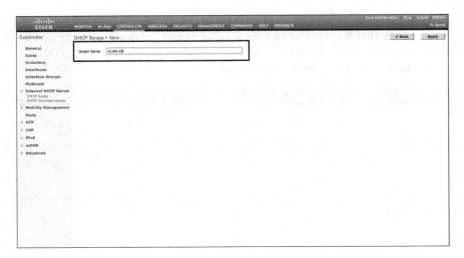

Figure 23-21   Naming a New DHCP Scope

You are returned to the DHCP Scopes page, shown in Figure 23-22, where you see your new scope appear in the list. It has been assigned an IP address pool of 0.0.0.0 – 0.0.0.0, it has a default lease time of one day, and its status is disabled.

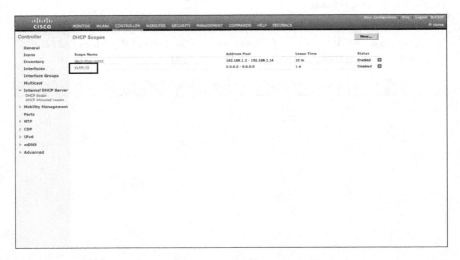

Figure 23-22   DHCP Scopes Page with New Scope Listed

To change the default 0.0.0.0 settings, click the name of the DHCP scope. You are taken to a new page, shown in Figure 23-23, where you can adjust your settings. Enter the necessary information such as pool start and end addresses, network, netmask, lease time, default routers, DNS servers, and NetBIOS name servers. Make sure to change the Status field to **Enabled**. Click the **Apply** button when finished, as indicated in Figure 23-23.

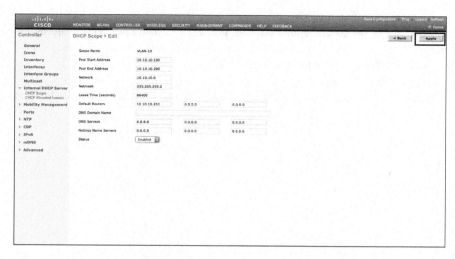

Figure 23-23   DHCP Scope Information Added

At this point you may wish to save your configuration so that it will be retained in the event of a loss of power. Click **Save Configuration** in the upper-right corner of the page. You will be asked to confirm this action, as shown in Figure 23-24. Click **OK**.

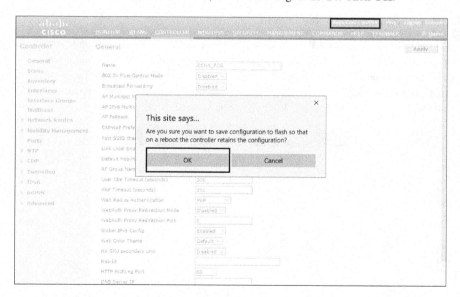

Figure 23-24   Saving Your Configuration

Configuring a WLAN

Depending on your needs, you may have a large or small number of WLANs. This section shows the steps needed to configure a WLAN on your WLC.

From the Advanced Monitor Summary screen, click **WLANs** in the top menu bar. You will see a list of already configured WLANs. Figure 23-25 shows one WLAN already created, the Staff_Network WLAN that was created previously with the simplified setup wizard. Click the **Go** button to create a new WLAN.

Figure 23-25   Creating a New WLAN

On the next screen, choose **WLAN** from the Type drop-down menu, enter the profile name and SSID, and choose your ID. The typical configuration, but not required, is to have the same profile name and SSID. Figure 23-26 shows this completed page, using 10 as the ID, to match with VLAN 10 that was created previously. Your choices for ID number range from 1 to 512. Click **Apply** when finished.

Figure 23-26   New WLAN Created

The next screen shows you what you entered on the last screen. Confirm it is correct. You also need to check the **Enabled** check box for this new WLAN, as shown in Figure 23-27.

NOTE: If you do not enable the WLAN, you will not be able to join the WLC from your wireless client.

Figure 23-27   Enabling the New WLAN

Defining a RADIUS Server

To authenticate wireless clients that wish to join your WLANs, you should define at least one RADIUS server to be used by your WLC.

From the Advanced Monitor Summary screen, click the **Security** tab to go to the Security General page, as shown in Figure 23-28. In the navigation pane on the left side of the screen, expand **AAA > RADIUS** to see the options for defining a RADIUS server.

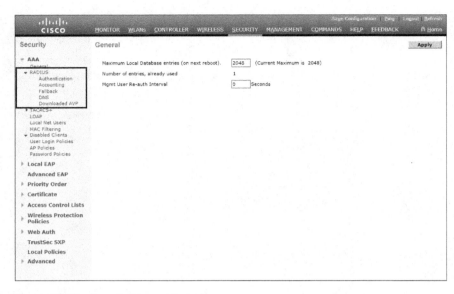

Figure 23-28 Security General Screen

Click **Authentication** to open the RADIUS Authentication Servers screen, shown in Figure 23-29, and click the **New** button to add a new RADIUS server.

Figure 23-29 Click New to Add a New RADIUS Server

Enter the required information about your RADIUS server, such as the IP address and your shared secret password, as shown in Figure 23-30. Click **Apply** when finished.

NOTE: Black dots are displayed instead of characters when you type your password in the Shared Secret and Confirm Shared Secret fields. Type carefully, because these passwords are case sensitive.

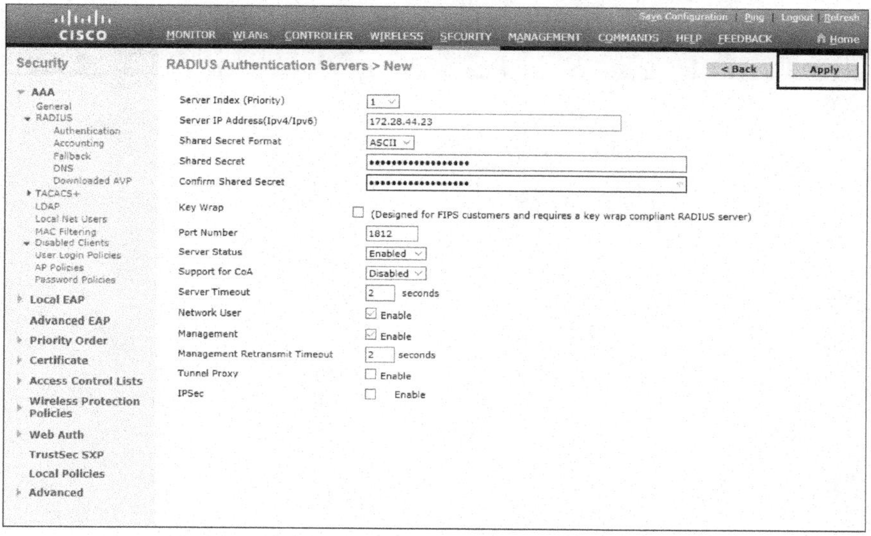

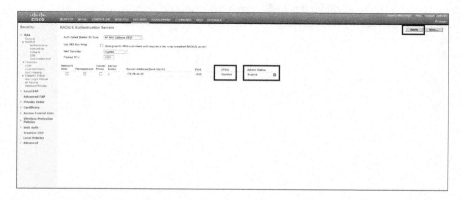

Figure 23-30   Adding Information About New RADIUS Server

The next screen shows you what you entered in the previous screen and gives you a chance to verify your information. Notice in Figure 23-31 that IPSec is set to Disabled and Admin Status is Enabled. Click **Apply** to continue.

Figure 23-31   RADIUS Authentication Servers Page

Exploring Management Options

The WLC can assist you in the management of your WLANs and in the management of your WLCs.

Figure 23-32 shows the Management Summary page, with the left navigation pane displaying the options that you can configure to help manage your network.

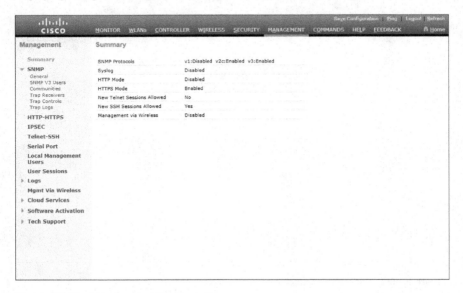

Figure 23-32   Management Summary Page

If you have multiple WLCs spread out across multiple locations, you can use the Simple Network Management Protocol (SNMP) and a centralized management console to manage your network. Figure 23-33 shows the SNMP System Summary page, which you reach by clicking **General** under SNMP in the navigation pane. Enter the name of your WLC, followed by its location and a contact name. Click **Apply** when you are finished.

NOTE: The name that is there to start is the name you gave to your WLC during the Simplified System Configuration wizard. If you do not wish to change the name, leave it as it is.

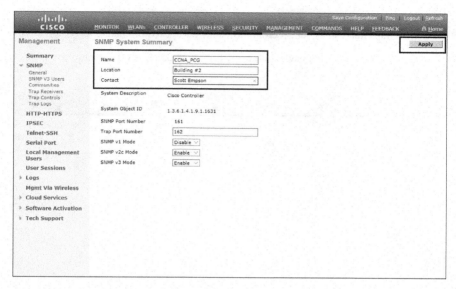

Figure 23-33 SNMP System Summary Page

To select which trap notifications you will receive, click **Trap Controls** under SNMP in the navigation pane; Figure 23-34 shows the SNMP Trap Controls with the General tab displayed.

TIP: Use the default settings to start, and adjust the traps to meet your needs.

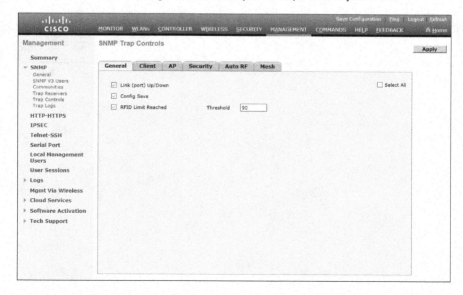

Figure 23-34 SNMP Trap Controls General Tab

Each of the other tabs—Client, AP, Security, Auto RF, and Mesh—all contain a list of the SNMP traps that can be set according to your network requirements.

Working your way down the navigation pane, Figure 23-35 shows the options available for HTTP-HTTPS configuration.

Figure 23-35   HTTP-HTTPS Configuration Page

Figure 23-36 shows the options available for Telnet-SSH configuration.

Figure 23-36   Telnet-SSH Configuration

Figure 23-37 shows the page that appears when you choose **Logs > Config**. Here you can set the IP address of the syslog server to which to send syslog messages. You can set up to three different syslog servers.

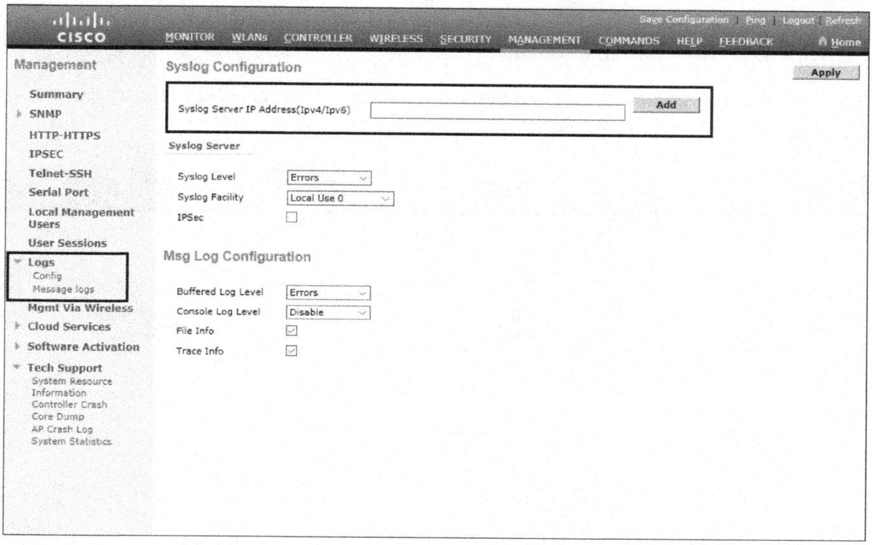

Figure 23-37 Syslog Configuration Page

Figure 23-38 shows the **Tech Support > System Resource Information** page, with settings for Current CPU Usage, Individual CPU Usage (shown only if you have multiple CPUs in your WLC), System Buffers, and Web Server Buffers.

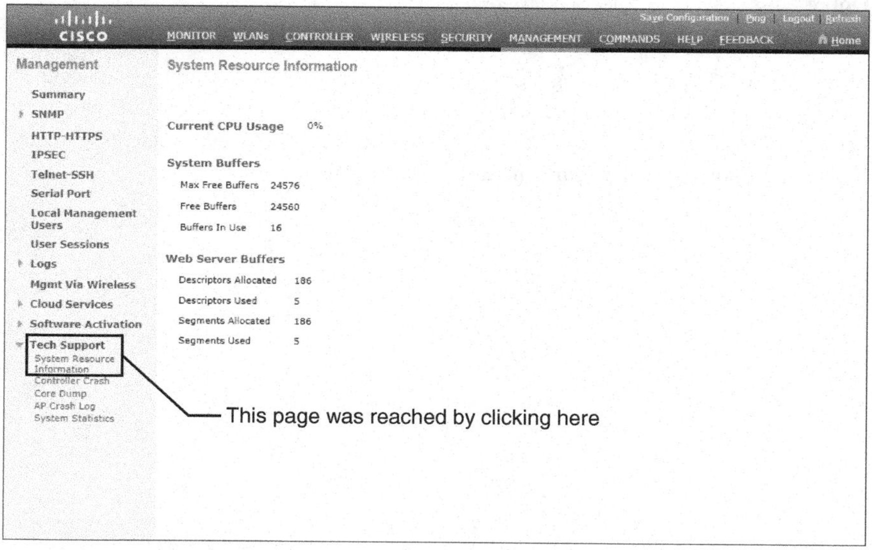

Figure 23-38 System Resource Information Page

Configuring a WLAN Using WPA2 PSK

WPA2 creates a framework for authentication and encryption. WPA2 has two modes of wireless protected access: WPA2 Personal mode, which uses WPA2-PSK; and WPA2 Enterprise mode, which uses IEEE 802.1X and EAP. This section shows you how you configure WPA2-PSK using the GUI.

Starting from the Advanced Monitor Summary screen, click **WLANs** in the top menu bar to see a screen that looks like Figure 23-39.

Figure 23-39   WLANs Main Screen

Create a new WLAN by clicking the **Go** button next to the drop-down menu showing Create New. This takes you to the screen shown in Figure 23-40, where you enter your WLAN information in the four fields: Type (leave as WLAN), Profile Name, SSID, and ID. Click **Apply** to commit your configuration to the WLC.

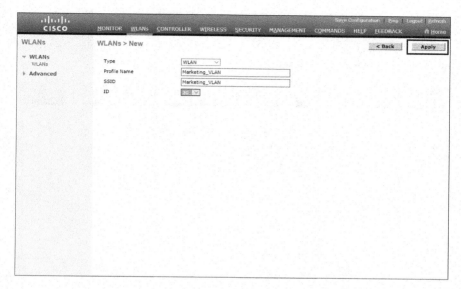

Figure 23-40   Creating and Applying a New WLAN

In the next screen, ensure that the status of your WLAN is Enabled, as shown in Figure 23-41.

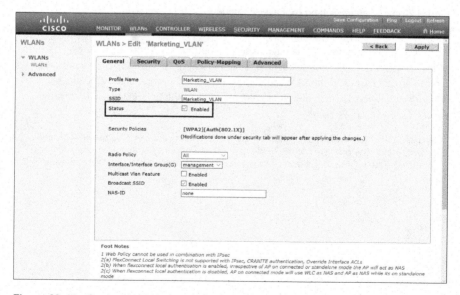

Figure 23-41   Setting the Status of the WLAN to Enabled

Next, move to the **Security** tab, shown in Figure 23-42. Notice the Layer 2 Security setting has defaulted to WPA+WPA2. Also notice that the default settings in the WPA+WPA2 Parameters section are to have the WPA2 Policy check box checked and the WPA2 Encryption option set to AES.

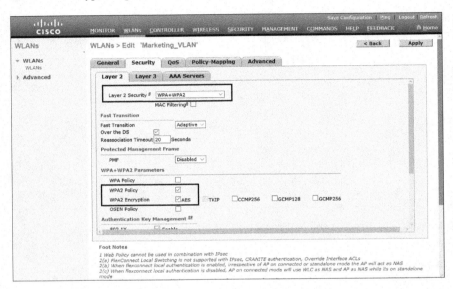

Figure 23-42   Default Setting for Layer 2 Security

Scroll down to the bottom of the page to see the options for Authentication Key Management. Click the **Enable** check box next to **PSK**. This action also unchecks the Enable check box next to 802.1X and displays new PSK Format options. Leave the default setting, ASCII, and enter your credential key in the field below it, as shown in Figure 23-43.

NOTE: When you type your credential key into the field, you see only black dots, and not your text. The right end of the field has a small symbol of an eye that you can click to see your text. Click it again to change back to black dots. Enter your text carefully.

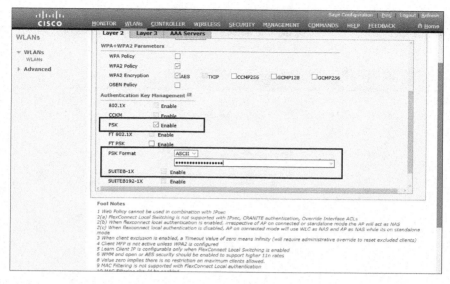

Figure 23-43   Authentication Key Management

When you are satisfied with your configuration, scroll back to the top of the page and click **Apply**. The warning popup box shown in Figure 23-44 appears, reminding you that changing parameters will cause your WLAN to be momentarily disabled and that there may be a loss of connectivity for some clients. Click **OK** to continue.

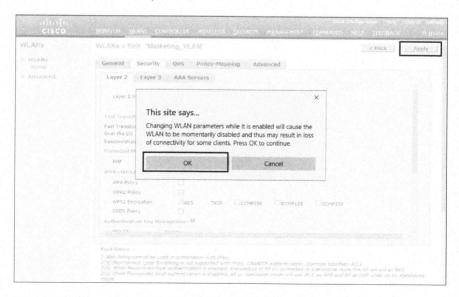

Figure 23-44   Applying and Confirming Your Configuration

After applying your configuration changes, you are returned to the WLANs General tab for this specific WLAN. Notice that the Security Policies field now indicates that you are running WPA2 with PSK authentication, as shown in Figure 23-45.

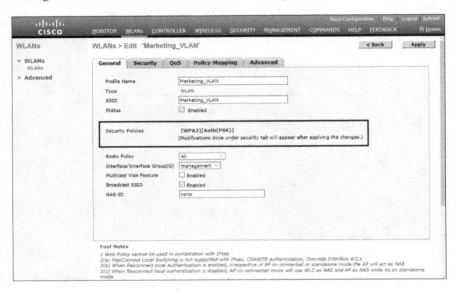

Figure 23-45   WLANs General Tab Showing Security Policies

How to Count in Decimal, Binary, and Hexadecimal

This appendix provides information concerning the following topics:

- How to count in decimal
- How to count in binary
- How to count in hexadecimal
- Representing decimal, binary, and hexadecimal numbers

So why does a book on Cisco IOS commands have a section on how to count? It may seem a bit strange or even redundant to have an appendix explaining how to count in decimal, but the fact is that the rules we all know for counting in decimal are the same rules we use to count in any other positional numbering system such as binary or hexadecimal. Because we all learned these rules when we were very young, we probably can't recite the rules; we just know them. But by reviewing the rules, you will be better prepared to learn how to work with binary and hexadecimal.

NOTE: I tell my students that when you know something so well that you no longer need to recite the rules to understand what is going on, you have *internalized* that information. You likely have internalized many things, like your phone number, how to drive a car, and the basic rules of writing a sentence. Something becomes internalized after much practice. With practice, you can also internalize working with binary and hexadecimal; just remember that you have been counting since you were very young. You have years of practice counting in decimal; binary and hex will come to you easily, after practice.

How to Count in Decimal

The decimal number system uses 10 as its base (and thus is also known as the base-10 system). It is the most widely used number system in modern civilizations. Ten unique symbols (or digits) represent different positions in the system: 0, 1, 2, 3, 4, 5, 6, 7, 8, and 9. Using these ten digits, we can represent any number that we want.

Using these ten digits, we start counting from 0 and get as far as

0

1

2

3

4

5

6

7

8

9

But what happens next? We have run out of unique digits, so we need to combine the digits to create larger numbers. This is where the idea of *positions* comes into play. Positional notation in decimal uses positions for each power of ten. The position of each digit within that number denotes the multiplier multiplied by that digit. This first column of numbers is located in what we call the *ones* position (or more correctly the 10^0 position). We then need to start the numbers over again, only now we insert numbers into the *tens* column (or the 10^1 column). Now we get this:

10	=	(1 multiplied by 10) plus (0 multiplied by 1)	=	$(1 \times 10) + (0 \times 1)$	=	10
11	=	(1 multiplied by 10) plus (1 multiplied by 1)	=	$(1 \times 10) + (1 \times 1)$	=	11
12	=	(1 multiplied by 10) plus (2 multiplied by 1)	=	$(1 \times 10) + (2 \times 1)$	=	12
13	=	(1 multiplied by 10) plus (3 multiplied by 1)	=	$(1 \times 10) + (3 \times 1)$	=	13
14	=	(1 multiplied by 10) plus (4 multiplied by 1)	=	$(1 \times 10) + (4 \times 1)$	=	14
15	=	(1 multiplied by 10) plus (5 multiplied by 1)	=	$(1 \times 10) + (5 \times 1)$	=	15
16	=	(1 multiplied by 10) plus (6 multiplied by 1)	=	$(1 \times 10) + (6 \times 1)$	=	16
17	=	(1 multiplied by 10) plus (7 multiplied by 1)	=	$(1 \times 10) + (7 \times 1)$	=	17
18	=	(1 multiplied by 10) plus (8 multiplied by 1)	=	$(1 \times 10) + (8 \times 1)$	=	18
19	=	(1 multiplied by 10) plus (9 multiplied by 1)	=	$(1 \times 10) + (9 \times 1)$	=	19

Because we have run out of numbers in the 1 × pattern (where × = a digit from 1 to 9), we then continue with the number 2 in the *tens* column:

20	=	(2 multiplied by 10) plus (0 multiplied by 1)	=	$(2 \times 10) + (0 \times 1)$	=	20
21	=	(2 multiplied by 10) plus (1 multiplied by 1)	=	$(2 \times 10) + (1 \times 1)$	=	21
22	=	(2 multiplied by 10) plus (2 multiplied by 1)	=	$(2 \times 10) + (2 \times 1)$	=	22
...						

We continue on until we reach the end of unique digits in the *tens* column, that being 99. After this we continue the pattern, only now we add the *hundreds* column (or the 10^2 column):

99	=	$(9 \times 10) + (9 \times 1)$	=	99
100	=	$(1 \times 100) + (0 \times 10) + (0 \times 1)$	=	100
101	=	$(1 \times 100) + (0 \times 10) + (1 \times 1)$	=	101
102	=	$(1 \times 100) + (0 \times 10) + (2 \times 1)$	=	102
...				
255	=	$(2 \times 100) + (5 \times 10) + (5 \times 1)$	=	255
...				
999	=	$(9 \times 100) + (9 \times 10) + (9 \times 1)$	=	999
1000	=	$(1 \times 1\,000) + (0 \times 100) + (0 \times 10) + (0 \times 1)$	=	1000
...				

Using the pattern of repetition, we continue to add digits to the columns as needed. Each column to the left grows by a power of 10:

...	10^7	10^6	10^5	10^4	10^3	10^2	10^1	10^0
	10 000 000	1 000 000	100 000	10 000	1 000	100	10	1

How to Count in Binary

The binary numeral system is a number system that uses 2 as its base (and thus is also known as the base-2 system). Because of its use in digital electronic circuitry using logic gates, binary is the number system used by all modern computers and computer-based devices. Two unique symbols (or digits) represent different positions in the system: 0 and 1. Using these two digits, we can represent any number that we want. Each digit is referred to as a *bit*.

Using these two digits, we start counting and get as far as:

0

1

So what happens next? We use the same rules that we used in decimal, only now each position uses a base of 2 instead of a base of 10:

...	2^{10}	2^9	2^8	2^7	2^6	2^5	2^4	2^3	2^2	2^1	2^0
	1024	512	256	128	64	32	16	8	4	2	1

The binary number system then looks like this:

Binary #					Decimal #
0	=	(0×2^0)	=	0	=	0
1	=	(1×2^0)	=	1	=	1
10	=	$(1 \times 2^1) + (0 \times 2^0)$	=	2 + 0	=	2
11	=	$(1 \times 2^1) + (1 \times 2^0)$	=	2 + 1	=	3
100	=	$(1 \times 2^2) + (0 \times 2^1) + (0 \times 2^0)$	=	4 + 0 + 0	=	4
101	=	$(1 \times 2^2) + (0 \times 2^1) + (1 \times 2^0)$	=	4 + 0 + 1	=	5
110	=	$(1 \times 2^2) + (1 \times 2^1) + (0 \times 2^0)$	=	4 + 2 + 0	=	6
111	=	$(1 \times 2^2) + (1 \times 2^1) + (1 \times 2^0)$	=	4 + 2 + 1	=	7
1000	=	$(1 \times 2^3) + (0 \times 2^2) + (0 \times 2^1) + (0 \times 2^0)$	=	8 + 0 + 0 + 0	=	8
1001	=	$(1 \times 2^3) + (0 \times 2^2) + (0 \times 2^1) + (1 \times 2^0)$	=	8 + 0 + 0 + 1	=	9
1010	=	$(1 \times 2^3) + (0 \times 2^2) + (1 \times 2^1) + (0 \times 2^0)$	=	8 + 0 + 2 + 0	=	10
1011	=	$(1 \times 2^3) + (0 \times 2^2) + (1 \times 2^1) + (1 \times 2^0)$	=	8 + 0 +2 + 1	=	11
1100	=	$(1 \times 2^3) + (1 \times 2^2) + (0 \times 2^1) + (0 \times 2^0)$	=	8 + 4 + 0 + 0	=	12
1101	=	$(1 \times 2^3) + (1 \times 2^2) + (0 \times 2^1) + (1 \times 2^0)$	=	8 + 4 + 0 + 1	=	13
1110	=	$(1 \times 2^3) + (1 \times 2^2) + (1 \times 2^1) + (0 \times 2^0)$	=	8 + 4 + 2 + 0	=	14
1111	=	$(1 \times 2^3) + (1 \times 2^2) + (1 \times 2^1) + (1 \times 2^0)$	=	8 + 4 + 2 + 1	=	15
10000	=	$(1 \times 2^4) + (0 \times 2^3) + (0 \times 2^2) + (0 \times 2^1)$ $+ (0 \times 2^0)$	=	16 + 0 + 0 $+ 0 + 0$	=	16
...					

How to Count in Hexadecimal

The hexadecimal (or hex) number system uses 16 as its base (and thus is also known as the base-16 system). It is used by computer systems designers and programmers. There are 16 unique symbols (or digits) to represent different positions in the system: 0 to 9, and then A to F.

Using the letters of A–F for numbers can be tricky at first, but it gets easier with practice. To help you out:

A in decimal = 10

B in decimal = 11

C in decimal = 12

D in decimal = 13

E in decimal = 14

F in decimal = 15

Each hex digit represents 4 bits (binary digits) and hex is often used to be a more human-friendly representation of large binary numbers.

To count in hex we use the same rules as decimal (base 10) and binary (base 2), except now we use a base of 16:

...	16^8	16^7	16^6	16^5	16^4	16^3	16^2	16^1	16^0
	4 294 967 296	268 435 456	16 777 216	1 048 576	65 536	4096	256	16	1

The following are some points of math trivia to help you when counting in hex:

16^2 is the same number as 2^8, which is the number of total (not valid) hosts in an IPv4 Class C network using a default subnet mask.

16^4 is the same number as 2^{16}, which is the number of total (not valid) hosts in an IPv4 Class B network using a default subnet mask.

16^6 is the same number as 2^{24}, which is the number of total (not valid) hosts in an IPv4 Class A network using a default subnet mask.

16^8 is the same number as 2^{32}, which is the number of total possible combinations in a 32-bit field, or the entire IPv4 addressing scheme.

16^{32} is the same number as 2^{128}, which is the number of total possible combinations in a 128-bit field, or the entire IPv6 addressing scheme.

16^{32}	=	2^{128}	=	340 282 366 920 938 463 463 374 607 431 768 211 456	=	3.4×10^{38}	=	340 undecillion

These are big, long numbers. Hence the reason to try and shorten them!

Counting in hex looks like this:

Hex #						Dec #
0	=	(0×16^0)	=	0×1	=	0
1	=	(1×16^0)	=	1×1	=	1
2	=	(2×16^0)	=	2×1	=	2
3	=	(3×16^0)	=	3×1	=	3
4	=	(4×16^0)	=	4×1	=	4
5	=	(5×16^0)	=	5×1	=	5
6	=	(6×16^0)	=	6×1	=	6
7	=	(7×16^0)	=	7×1	=	7
8	=	(8×16^0)	=	8×1	=	8
9	=	(9×16^0)	=	9×1	=	9
A	=	$(A \times 16^0)$	=	10×1	=	10
B	=	$(B \times 16^0)$	=	11×1	=	11
C	=	$(C \times 16^0)$	=	12×1	=	12
D	=	$(D \times 16^0)$	=	13×1	=	13

Hex #							Dec #
E	=	$(E \times 16^0)$	=	14×1	=		14
F	=	$(F \times 16^0)$	=	15×1	=		15
10	=	$(1 \times 16^1) + (0 \times 16^0)$	=	$(1 \times 16) + (0 \times 1)$	=		16
11		$(1 \times 16^1) + (1 \times 16^0)$	=	$(1 \times 16) + (1 \times 1)$	=		17
12		$(1 \times 16^1) + (2 \times 16^0)$	=	$(1 \times 16) + (2 \times 1)$	=		18
13		$(1 \times 16^1) + (3 \times 16^0)$	=	$(1 \times 16) + (3 \times 1)$	=		19
...							
1F	=	$(1 \times 16^1) + (F \times 16^0)$	=	$(1 \times 16) + (15 \times 1)$	=		31
20	=	$(2 \times 16^1) + (0 \times 16^0)$	=	$(2 \times 16) + (0 \times 1)$	=		32
21	=	$(2 \times 16^1) + (1 \times 16^0)$	=	$(2 \times 16) + (1 \times 1)$	=		33
...							
2F	=	$(2 \times 16^1) + (F \times 16^0)$	=	$(2 \times 16) + (15 \times 1)$	=		47
30	=	$(3 \times 16^1) + (0 \times 16^0)$	=	$(3 \times 16) + (0 \times 1)$	=		48
...							
64	=	$(6 \times 16^1) + (4 \times 16^0)$	=	$(6 \times 16) + (4 \times 1)$	=		100
...							
FF	=	$(F \times 16^1) + (F \times 16^0)$	=	$(15 \times 16) + (15 \times 1)$	=		255
100	=	$(1 \times 16^2) + (0 \times 16^1) + (0 \times 16^0)$	=	$(256) + (0) + (0)$	=		256
...							
FACE	=	$(F \times 16^3) + (A \times 16^2) + (C \times 16^1) + (E \times 16^0)$	=	$(61440) + (2560) + (192) + (14)$	=		64 206
...							

Representing Decimal, Binary, and Hexadecimal Numbers

In a perfect world, one numbering system would be used for everything and there would be no confusion. But the world isn't perfect, so we have many different number systems, and we must distinguish between them.

- **Decimal:** Because this is the system used most often, assume that any written number is a decimal number. The proper way to denote decimal numbers is to use a subscript of DEC or 10 after the number:

  $100 = 100_{DEC} = 100_{10}$

- **Binary:** To represent binary digits, append a subscript of either BIN or 2:

  $10111001_{BIN} = 10111001_2$

However, when using binary in IP addressing, seeing a group of 1s and 0s in a long line is a pretty good indicator that binary is being used. Seeing a number that is broken down into groups of eight is also a good indicator that binary is being used:

111111111111000000000000000000000000 = 11111111 11110000 00000000 00000000

- **Hexadecimal:** To represent hex digits, append a subscript of HEX or use a prefix of 0x:

$100_{HEX} = 0x100$

Seeing a number with letters in it (A–F) is also a sure bet that hex is being used:

$2001ABCD = 0x2001ABCD = 2001ABCD_{HEX}$

How to Convert Between Number Systems

This appendix provides information concerning the following topics:

- How to convert from decimal to binary
- How to convert from binary to decimal
- How to convert from decimal IP addresses to binary and binary IP addresses to decimal
- How to convert from hexadecimal to binary
- How to convert from binary to hexadecimal
- How to convert from decimal to hexadecimal
- How to convert from hexadecimal to decimal

Now that we have covered how to count (in Appendix A), we need to be able to convert between the three different number systems. This is another skill that takes time and practice to become comfortable using, and it is a skill that can quickly be lost without usage.

How to Convert from Decimal to Binary

Remember that in the binary number system, we use 2 as our base number, giving us a chart that looks like this:

...	2^{10}	2^9	2^8	2^7	2^6	2^5	2^4	2^3	2^2	2^1	2^0
	1 024	512	256	128	64	32	16	8	4	2	1

> **TIP:** Re-create this chart right before taking an exam. It is quick to write out on a piece of paper (or the erasable board you get in a vendor exam) and will help you with your addition. I have seen many students lose marks because of simple addition errors.

Each of these numbers represents a bit that has been turned on; that is, represented by a 1 in a bit pattern. So, to convert a decimal number to binary, you must add up these numbers to equal the decimal number—those numbers are then turned on and converted to 1 in the pattern:

1200	=	1024 + 128 + 32 + 16	=	10010110000
755	=	512 + 128 + 64 + 32 + 16 + 2 + 1	=	1011110011
500	=	256 + 128 + 64 + 32 + 16 + 4	=	111110100
233	=	128 + 64 + 32 + 8 + 1	=	11101001
187	=	128 + 32 + 16 + 8 + 2 + 1	=	10111011
160	=	128 + 32	=	10100000
95	=	64 + 16 + 8 + 4 + 2 + 1	=	1011111
27	=	16 + 8 + 2 + 1	=	11011

In IPv4, we use 32 bits to represent an IP address. These 32 bits are broken down into four groups of 8 bits. Each group is called an *octet*. You should become very comfortable with working with octets and being able to convert decimal into octets and vice versa. When working with octets, it is customary to insert leading 0s in their placeholders so that all positions in the octet are represented:

| 95 | = | 64 + 16 + 8 + 4 + 2 + 1 | = | 01011111 |
| 27 | = | 16 + 8 + 2 + 1 | = | 00011011 |

Some numbers are used over and over again in IPv4 addresses, so these conversions will be easy to remember:

		2^7	2^6	2^5	2^4	2^3	2^2	2^1	2^0		
Decimal #		128	64	32	16	8	4	2	1	=	Binary #
0	=	0	0	0	0	0	0	0	0	=	00000000
1	=	0	0	0	0	0	0	0	1	=	00000001
64	=	0	1	0	0	0	0	0	0	=	01000000
128	=	1	0	0	0	0	0	0	0	=	10000000
192	=	1	1	0	0	0	0	0	0	=	11000000
224	=	1	1	1	0	0	0	0	0	=	11100000
240	=	1	1	1	1	0	0	0	0	=	11110000
248	=	1	1	1	1	1	0	0	0	=	11111000
252	=	1	1	1	1	1	1	0	0	=	11111100
254	=	1	1	1	1	1	1	1	0	=	11111110
255	=	1	1	1	1	1	1	1	1		11111111

How to Convert from Binary to Decimal

When converting binary numbers to decimal, we need to use the base-2 chart just like when converting decimal to binary:

...	2^{10}	2^9	2^8	2^7	2^6	2^5	2^4	2^3	2^2	2^1	2^0
	1 024	512	256	128	64	32	16	8	4	2	1

TIP: Re-create this chart right before taking an exam. It is quick to write out on a piece of paper (or the erasable board you get in a vendor exam) and will help you with your addition. I have seen many students lose marks because of simple addition errors.

Using this chart as our guide, we add up all the numbers that are represented by a 1 in the binary pattern to convert it to a decimal number:

101011110000	=	1024 + 256 + 64 + 32 + 16 + 8	=	1400
1011110011	=	512 + 128 + 64 + 32 + 16 + 2 + 1	=	755
111100010	=	256 + 128 + 64 + 32 + 2	=	482

11100100	=	128 + 64 + 32 + 4	=	228
10111010	=	128 + 32 + 16 + 8 + 2	=	186
10101010	=	128 + 32 + 8 + 2	=	170
1001101	=	64 + 8 + 4 + 1	=	77
1101	=	8 + 4 + 1	=	13

Again, we use leading 0s in order to form octets:

| 01001101 | = | 64 + 8 + 4 + 1 | = | 77 |
| 00001101 | = | 8 + 4 + 1 | = | 13 |

How to Convert from Decimal IP Addresses to Binary and from Binary IP Addresses to Decimal

To convert a decimal IP address to binary, we need to take each individual number and convert it to binary:

192.168.10.1				
	192	=	11000000	
	168	=	10101000	
	10	=	00001010	
	1	=	00000001	
			=	11000000 10101000 00001010 00000001
172.16.100.254				
	172	=	10101100	
	16	=	00010000	
	100	=	01100100	
	254	=	11111110	
			=	10101100 00010000 01100100 11111110

TIP: When writing out IP addresses in binary, it is customary to put either a space or a period between the octets:

10101100 00010000 01100100 11111110 *or* 10101100.00010000.01100100.11111110

To convert a binary IP address to decimal, we need to take each individual binary number and convert it to decimal:

11000000 10101000 01100100 11110011					
	11000000	=	192		
	10101000	=	168		
	01100100	=	100		
	1111011	=	243		
				=	192.168.100.243
00001010 10010110 01100000 00000001					
	00001010	=	10		
	10010110	=	150		
	01100000	=	96		
	00000001	=	1		
				=	10.150.96.1

A Bit of Perspective

In the IPv4 addressing scheme, there are 2^{32} potential unique addresses that are mathematically possible. If we didn't use the dotted-decimal format of breaking down 32-bit address into four groups of 8 bits, we may have to use numbers like 3,232,238,081 for 192.168.10.1, or 84,215,041 instead of 10.10.10.1. Which would be easier for you to remember?

How to Convert from Hexadecimal to Binary

This is a conversion that you will be using more than converting hex to decimal. It is very simple to do. Each hex digit can be represented by four binary digits using the following table:

Hex	Binary
0	0000
1	0001
2	0010
3	0011
4	0100
5	0101
6	0110
7	0111
8	1000
9	1001
A	1010

B	1011
C	1100
D	1101
E	1110
F	1111

TIP: Re-create this chart right before taking an exam. It is quick to write out on a piece of paper (or the erasable board you get in a vendor exam) and will help you with your conversions. I have seen many students lose marks because of simple errors.

When converting hex to binary, it is customary to convert pairs of hex digits to binary octets, as shown in Example 1.

Example 1: Convert C9 to Binary

Step 1	C	=	1100
Step 2	9	=	1001
Therefore	C9	=	11001001

If there is an uneven number of hex digits to make a complete octet, put a leading 0 at the *beginning* of the hex number, as shown in Example 2.

Example 2: Convert DB8 to Binary

Step 1	D	=	1101
Step 2	B	=	1011
Step 3	8	=	1000
Therefore	DB8	=	00001101 10111000

NOTE: Leading 0s are used to complete the first octet (DB8 = 0DB8)

How to Convert from Binary to Hexadecimal

This is also a conversion that you will be using more than converting decimal to hex. Using the same table as shown in the preceding section, convert each octet of binary to a pair of hex digits, as shown in Example 3.

Example 3: Convert 11000101 to Hex

Step 1	1100	=	C
Step 2	0101	=	5
Therefore	11000101	=	C5

If there is an uneven number of binary digits to make a complete octet, put one or more leading 0s at the *beginning* of the binary number, as shown in Example 4.

Example 4: Convert 111111110101 to Binary

Step 1	1111	=	F
Step 2	1111	=	F
Step 3	0101	=	5
Step 4	Because there aren't enough hex digits to make a complete pair, add leading 0s to complete the pair. In this case, add 0000 to the beginning of the number.		
Therefore	0000111111110101	=	0F F5

> **NOTE:** Leading 0s are used to complete the first octet (0FF5 = FF5)

> **TIP:** Although it takes an extra step, I find it faster to convert hexadecimal digits to binary and then binary to decimal. I can add and subtract faster than multiplying and dividing by 16s.

Example 5 shows the process for converting an entire IP address from hex to decimal.

Example 5: Convert the IP Address C0A80101$_{HEX}$ to Decimal

Step 1	C0	=	11000000	=	192
Step 2	A8	=	10101000	=	168
Step 3	01	=	00000001	=	1
Step 4	01	=	00000001	=	1
Therefore	C0A80101$_{HEX}$	=			192.168.1.1

IPv4 addresses are not usually represented in hexadecimal; an exception is when you are capturing traffic with software such as Wireshark, which shows all traffic in hex.

How to Convert from Decimal to Hexadecimal

Although this type of conversion is not used very often, you should still understand how to convert a decimal number to a hexadecimal number. To do this conversion, you have to use long division while leaving the remainders in integer form—do not use decimal answers!

The steps involved are as follows:

1. Divide the decimal number by 16. Leave the remainder in integer form (*no decimals*).

   Write down the remainder in hexadecimal form.

2. Divide the result from step 1 by 16.

   Write down the remainder in hexadecimal form.

3. Repeat this process until you are left with an answer of 0.

   If there is a remainder, write that down in hexadecimal form.

4. The hex value is the sequence of remainders from the last one to the first one.

Examples 6, 7, and 8 show how to convert decimal numbers to hexadecimal.

Example 6: Convert 188_{10} to Hex

Step 1	188 / 16	=	11 with a remainder of 12	12 in hex is C
Step 2	11 / 16	=	0 with a remainder of 11	11 in hex is B
Therefore	188_{10}	=		BC_{HEX}

Example 7: Convert 255_{10} to Hex

Step 1	255 / 16	=	15 with a remainder of 15	15 in hex is F
Step 2	15 / 16	=	0 with a remainder of 15	15 in hex is F
Therefore	255_{10}	=		FF_{HEX}

Example 8: Convert 1234_{10} to Hex

Step 1	1234 / 16	=	77 with a remainder of 2	2 in hex is 2
Step 2	77 / 16	=	4 with a remainder of 13	13 in hex is D
Step 3	4 / 16	=	0 with a remainder of 4	4 in hex is 4
Therefore	1234_{10}	=		$4D2_{HEX}$

How to Convert from Hexadecimal to Decimal

This is also a type of conversion that isn't used very often. This time we will be multiplying by powers of 16 to convert hexadecimal numbers to decimal.

The steps involved are as follows:

1. Take the last digit (also known as the least significant digit) in the hexadecimal number and multiply it by 16^0. Put this number off to the side for right now.

   Remember that $16^0 = 1$, so you are multiplying this digit by 1.

2. Take the second-to-last digit in the hex number and multiply it by 16^1. Also put this number off to the side.

   You are multiplying this digit by 16.

3. Continue multiplying the digits with increasing powers of 16 until you are finished multiplying all of the individual hexadecimal digits.

4. Add up all of the results from your multiplications and you have your answer in decimal.

Examples 9 through 11 show you how to convert a hexadecimal number to decimal.

Example 9: Convert C3$_{\text{HEX}}$ to Decimal

Step 1	3×16^0	=	3×1	=	3
Step 2	$C \times 16^1$	=	12×16	=	192
Therefore	C3$_{\text{HEX}}$	=			195

Example 10: Convert 276$_{\text{HEX}}$ to Decimal

Step 1	6×16^0	=	6×1	=	6
Step 2	7×16^1	=	7×16	=	112
Step 3	2×16^2	=	2×256	=	512
Therefore	276$_{\text{HEX}}$	=			630

Example 11: Convert FACE$_{\text{HEX}}$ to Decimal

Step 1	$E \times 16^0$	=	14×1	=	14
Step 2	$C \times 16^1$	=	12×16	=	192
Step 3	$A \times 16^2$	=	10×256	=	2560
Step 4	$F \times 16^3$	=	15×4096	=	61 440
Therefore	FACE$_{\text{HEX}}$	=			64 206

Binary/Hex/Decimal Conversion Chart

The following chart lists the three most common number systems used in networking: decimal, hexadecimal, and binary. Some numbers you will remember quite easily, as you use them a lot in your day-to-day activities. For those other numbers, refer to this chart.

Decimal Value	Hexadecimal Value	Binary Value
0	00	0000 0000
1	01	0000 0001
2	02	0000 0010
3	03	0000 0011
4	04	0000 0100
5	05	0000 0101
6	06	0000 0110
7	07	0000 0111
8	08	0000 1000
9	09	0000 1001
10	0A	0000 1010
11	0B	0000 1011
12	0C	0000 1100
13	0D	0000 1101
14	0E	0000 1110
15	0F	0000 1111
16	10	0001 0000
17	11	0001 0001
18	12	0001 0010
19	13	0001 0011
20	14	0001 0100
21	15	0001 0101
22	16	0001 0110
23	17	0001 0111
24	18	0001 1000
25	19	0001 1001
26	1A	0001 1010
27	1B	0001 1011

Decimal Value	Hexadecimal Value	Binary Value
28	1C	0001 1100
29	1D	0001 1101
30	1E	0001 1110
31	1F	0001 1111
32	20	0010 0000
33	21	0010 0001
34	22	0010 0010
35	23	0010 0011
36	24	0010 0100
37	25	0010 0101
38	26	0010 0110
39	27	0010 0111
40	28	0010 1000
41	29	0010 1001
42	2A	0010 1010
43	2B	0010 1011
44	2C	0010 1100
45	2D	0010 1101
46	2E	0010 1110
47	2F	0010 1111
48	30	0011 0000
49	31	0011 0001
50	32	0011 0010
51	33	0011 0011
52	34	0011 0100
53	35	0011 0101
54	36	0011 0110
55	37	0011 0111
56	38	0011 1000
57	39	0011 1001
58	3A	0011 1010
59	3B	0011 1011
60	3C	0011 1100
61	3D	0011 1101
62	3E	0011 1110
63	3F	0011 1111
64	40	0100 0000

Decimal Value	Hexadecimal Value	Binary Value
65	41	0100 0001
66	42	0100 0010
67	43	0100 0011
68	44	0100 0100
69	45	0100 0101
70	46	0100 0110
71	47	0100 0111
72	48	0100 1000
73	49	0100 1001
74	4A	0100 1010
75	4B	0100 1011
76	4C	0100 1100
77	4D	0100 1101
78	4E	0100 1110
79	4F	0100 1111
80	50	0101 0000
81	51	0101 0001
82	52	0101 0010
83	53	0101 0011
84	54	0101 0100
85	55	0101 0101
86	56	0101 0110
87	57	0101 0111
88	58	0101 1000
89	59	0101 1001
90	5A	0101 1010
91	5B	0101 1011
92	5C	0101 1100
93	5D	0101 1101
94	5E	0101 1110
95	5F	0101 1111
96	60	0110 0000
97	61	0110 0001
98	62	0110 0010
99	63	0110 0011
100	64	0110 0100
101	65	0110 0101

Decimal Value	Hexadecimal Value	Binary Value
102	66	0110 0110
103	67	0110 0111
104	68	0110 1000
105	69	0110 1001
106	6A	0110 1010
107	6B	0110 1011
108	6C	0110 1100
109	6D	0110 1101
110	6E	0110 1110
111	6F	0110 1111
112	70	0111 0000
113	71	0111 0001
114	72	0111 0010
115	73	0111 0011
116	74	0111 0100
117	75	0111 0101
118	76	0111 0110
119	77	0111 0111
120	78	0111 1000
121	79	0111 1001
122	7A	0111 1010
123	7B	0111 1011
124	7C	0111 1100
125	7D	0111 1101
126	7E	0111 1110
127	7F	0111 1111
128	80	1000 0000
129	81	1000 0001
130	82	1000 0010
131	83	1000 0011
132	84	1000 0100
133	85	1000 0101
134	86	1000 0110
135	87	1000 0111
136	88	1000 1000
137	89	1000 1001
138	8A	1000 1010

Decimal Value	Hexadecimal Value	Binary Value
139	8B	1000 1011
140	8C	1000 1100
141	8D	1000 1101
142	8E	1000 1110
143	8F	1000 1111
144	90	1001 0000
145	91	1001 0001
146	92	1001 0010
147	93	1001 0011
148	94	1001 0100
149	95	1001 0101
150	96	1001 0110
151	97	1001 0111
152	98	1001 1000
153	99	1001 1001
154	9A	1001 1010
155	9B	1001 1011
156	9C	1001 1100
157	9D	1001 1101
158	9E	1001 1110
159	9F	1001 1111
160	A0	1010 0000
161	A1	1010 0001
162	A2	1010 0010
163	A3	1010 0011
164	A4	1010 0100
165	A5	1010 0101
166	A6	1010 0110
167	A7	1010 0111
168	A8	1010 1000
169	A9	1010 1001
170	AA	1010 1010
171	AB	1010 1011
172	AC	1010 1100
173	AD	1010 1101
174	AE	1010 1110
175	AF	1010 1111

Decimal Value	Hexadecimal Value	Binary Value
176	B0	1011 0000
177	B1	1011 0001
178	B2	1011 0010
179	B3	1011 0011
180	B4	1011 0100
181	B5	1011 0101
182	B6	1011 0110
183	B7	1011 0111
184	B8	1011 1000
185	B9	1011 1001
186	BA	1011 1010
187	BB	1011 1011
188	BC	1011 1100
189	BD	1011 1101
190	BE	1011 1110
191	BF	1011 1111
192	C0	1100 0000
193	C1	1100 0001
194	C2	1100 0010
195	C3	1100 0011
196	C4	1100 0100
197	C5	1100 0101
198	C6	1100 0110
199	C7	1100 0111
200	C8	1100 1000
201	C9	1100 1001
202	CA	1100 1010
203	CB	1100 1011
204	CC	1100 1100
205	CD	1100 1101
206	CE	1100 1110
207	CF	1100 1111
208	D0	1101 0000
209	D1	1101 0001
210	D2	1101 0010
211	D3	1101 0011
212	D4	1101 0100

Decimal Value	Hexadecimal Value	Binary Value
213	D5	1101 0101
214	D6	1101 0110
215	D7	1101 0111
216	D8	1101 1000
217	D9	1101 1001
218	DA	1101 1010
219	DB	1101 1011
220	DC	1101 1100
221	DD	1101 1101
222	DE	1101 1110
223	DF	1101 1111
224	E0	1110 0000
225	E1	1110 0001
226	E2	1110 0010
227	E3	1110 0011
228	E4	1110 0100
229	E5	1110 0101
230	E6	1110 0110
231	E7	1110 0111
232	E8	1110 1000
233	E9	1110 1001
234	EA	1110 1010
235	EB	1110 1011
236	EC	1110 1100
237	ED	1110 1101
238	EE	1110 1110
239	EF	1110 1111
240	F0	1111 0000
241	F1	1111 0001
242	F2	1111 0010
243	F3	1111 0011
244	F4	1111 0100
245	F5	1111 0101
246	F6	1111 0110
247	F7	1111 0111
248	F8	1111 1000
249	F9	1111 1001

Decimal Value	Hexadecimal Value	Binary Value
250	FA	1111 1010
251	FB	1111 1011
252	FC	1111 1100
253	FD	1111 1101
254	FE	1111 1110
255	FF	1111 1111

Create Your Own Journal Here

G

gigabit Ethernet interfaces, assigning IPv4 addresses, 132

global configuration mode, routers, 126

GUAs (global unicast addresses), 45–46

guidelines, for configuring EtherChannel, 112–113

H

hello interval timer, OSPF (Open Shortest Path First), 153

hello-time command, 102

hello-time keyword, 100

help
keyboard help, 62–63
question mark (?), 60–61

help commands, configuring switches, 68

hexidecimal digits, IPv6 addresses, 40

hierarchical addresses, IPv4 addresses, 1

history commands, 63

history size command, 64

host addresses, 3

host bits, 11

host keyword, 198

host names, setting for switches, 69

hostname command, 219

HTTP service, disabling, 221

HTTP-HTTPS Configuration page, 244

I

IEEE Standard 802.1Q (dot1q), 84

IETF (Internet Engineering Task Force), 39–40

illegal characters in host names, 69

implementing logging, 214

implicit deny rule, 211

in keyword, 200

information, verifying for VLANs, 78

interface descriptions, configuring, switches, 70

interface modes,
EtherChannel, 111
routers, 126

interface names, routers, 127–131

interface range command, 76

interfaces, moving between, 131

Internet Engineering Task Force (IETF), 39–40

Inter-Switch Link (ISL), 84

inter-VLAN communication
configuration examples, 89
CORP routers, 90–92
ISP router, 89–90
L2Switch1 (Catalyst 2960), 95–96
L2Switch2 (Catalyst 2960), 92–94
L3Switch1 (Catalyst 3560/3650/3750), 94–95
with external routers (router-on-a-stick), 87
on multilayer switches, through SVI (Switch Virtual Interface), 88
network topology, 89
tips for, 88–89

IOS routers, configuring DHCP servers, 159–160

IOS software ethernet interface, configuring DHCP clients, 162

ip access-list resequence command, 205

IP addresses, configuring switches, 70

ip forward-helper udp x, 161

ip helper-address, 161

ip name-server command, 135

ip ospf process ID area area number command, 151

IP plans, VLSM example, 24–31

IP redirects, disabling, 221

ip route, 141–142

IP source routing, disabling, 221

ip subnet zero, 23

IPv4 ACLs, configuration examples, 208–210

IPv4 addresses, 39–40
appearance of, 2